Agricultural Development in Asia - Potential Use of Nano-Materials and Nano-Technology

Edited by Md. Asaduzzaman and Mafruha Afroz

Published in London, United Kingdom

IntechOpen

Supporting open minds since 2005

Agricultural Development in Asia - Potential Use of Nano-Materials and Nano-Technology
http://dx.doi.org/10.5772/intechopen.78083
Edited by Md. Asaduzzaman and Mafruha Afroz

Contributors
Rahul Sadhukhan, L. Devarishi Sharma, Suman Sen, Snehashis Karmakar, Koushik Banerjee, Krittiranjan Baral, Muhammad Irfan Said, Wasan S. Hussain, Mahmoud Magdy Abbas, Saroj Kumar Yadav, Kiran P. Raverkar, Malik Asif Aziz, Zaffar Mahdi Dar, Mushtaq A. Malik, Saima Shafi, Bisma Rashid, Shayesta Islam, Showkat Sidique, Amjad Masood, Imteyaz Alam, Nusrat Iqbal, Amrish Agrawal, Jitendra Kumar, Md Asaduzzaman, Mohammad Monirul Hasan Tipu, Artho Baroi, Juwel Rana, Shariful Islam, Raunak Jahan, Md. Shipon Miah

First published in London, United Kingdom, 2022 by IntechOpen
IntechOpen is the global imprint of INTECHOPEN LIMITED, registered in England and Wales, registration number: 11086078, 5 Princes Gate Court, London, SW7 2QJ, United Kingdom
Printed in Croatia

British Library Cataloguing-in-Publication Data
A catalogue record for this book is available from the British Library

Additional hard and PDF copies can be obtained from orders@intechopen.com

Agricultural Development in Asia - Potential Use of Nano-Materials and Nano-Technology
Edited by Md. Asaduzzaman and Mafruha Afroz
p. cm.
Print ISBN 978-1-83968-136-3
Online ISBN 978-1-83968-137-0
eBook (PDF) ISBN 978-1-83968-138-7

For EU product safety concerns:
IN TECH d.o.o., Prolaz Marije Krucifikse Kozulić 3, 51000 Rijeka, Croatia, info@intechopen.com or visit our website at intechopen.com.

Meet the editors

Dr. Asaduzzaman is a native of Bangladesh and received a Ph.D. in Bioproduction Science from Tottori University, Japan. He has expertise in hydroponic crop production and is currently working as a senior researcher at the Horticulture Research Centre, Bangladesh Agricultural Research Institute. His main research focuses on the development of hydroponic techniques for horticultural crops in a greenhouse, production of specialty crops under Controlled Environment Agriculture (CEA), and development of specialty dietary components through hydroponic production of vegetables providing human health benefits beyond basic nutrition. His other research project includes studying autotoxicity, a phenomenon of intraspecific allelopathy in vegetables and ornamentals through hydroponics, and developing suitable control measures to overcome it. He has published thirty-one original research articles, five review articles, twenty-two conference proceedings, eight book chapters, and nine edited books. He was awarded the Gold Medal from Bangladesh Agricultural University in 2011 and the 2016 BAS-TWAS Prize for Young Scientists from Bangladesh.

Dr. Mafruha Afroz is a plant pathologist at the Bangladesh Agricultural Research Institute, Bangladesh, where she has worked for more than 15 years. Dr. Afroz received a Ph.D. in Bacteriology from Ohio State University, USA. Her main research interests are diagnosis and molecular characterization of plant pathogens and their appropriate management, and management of diseases using varietal resistance as well as cultural, chemical, and biological tools. She has published several original research articles in reputed national and international journals.

Contents

Preface

Agricultural development is a potential way to mitigate world hunger and achieve sustainable development. It is a lifelong process through which poor populations can achieve food security and improve their nutrition. Improvement of crop varietals along with technological innovation can lead towards a prosperous life. In recent years, nanomaterials and nanotechnology have been used to promote plant growth and yield in Asia.

This book reviews the potential applications of nanotechnology for sustainable agriculture. It examines the synthesis and use of nanofertilizers, nanopesticides, growth-promoting microorganisms, and allelochemicals in precision agriculture as well as their beneficial and detrimental impacts on the environment. The book also presents recent research on these materials.

We gratefully acknowledge the many researchers who contributed to this volume.

Md. Asaduzzaman
Olericulture Division,
Horticultural Research Centre,
Bangladesh Agricultural Research Institute,
Gazipur, Bangladesh

Mafruha Afroz
Plant Pathology Section,
Horticultural Research Centre,
Bangladesh Agricultural Research Institute,
Gazipur, Bangladesh

Chapter 1

Potential Applications of Nanotechnology in Agriculture: A Smart Tool for Sustainable Agriculture

Mohammad Monirul Hasan Tipu, Artho Baroi, Juwel Rana, Shariful Islam, Raunak Jahan, Md. Shipon Miah and Md. Asaduzzaman

Abstract

Most of the early uses of nanotechnology have come from material sciences, although applications in agriculture are still expanding. Due to a few comprehensive reviews, we described application of nanomaterials along with their fate in soil and interaction with soil and plant system. From synthesis to metabolism, nano-fertilizers like zinc, silver, selenium, titanium oxide have enhanced the physio-chemical characteristics of crop plants in every manner conceivable. On the other hand, it has the potential to minimize pesticide use by boosting reactivity and surface area of nanoparticles. Nanotechnology in pesticides will, without a doubt, replace the current way of pesticide application because of its efficacy. Nano-based approaches can readily overcome the constraints of conventional soil remediation technologies. While soil nanomaterials mobility has been investigated in a limited number of research studies, it's likely the most critical gap in knowing the real risk of their transport. As well as enhancing plant nutrient absorption, nanomaterials may also be used to regulate soil microbial activity and stimulate plant defenses. When it comes to shipping food, nanotechnology has made things easier by extending the shelf life of most foods. While it offers tremendous potential for agricultural applications, the health effects of nanoparticles on plants, animals, and humans must be thoroughly investigated.

Keywords: nanotechnology, nanomaterials, agriculture, pesticide, fertilizer, plant defense, soil remediation, agricultural waste

1. Introduction

Agriculture is the most important and reliable enterprise, as it provides food and raw materials for industries that require them. The depletion of natural resources and the increase in world population demand that agricultural production become more economically feasible, ecologically sound, and productive. Changes such as these are crucial to attaining a number of goals inside the last year. Therefore, environmental performance should be employed and involvement from food chain ecosystems concerning agricultural foodstuff production must be incorporated. When it comes to eliminating poverty and hunger from today's world, the agricultural revolution is a must-have phenomenon. In this scenario, well-known lenders are living below the poverty line and are dispersed in rural areas where agricultural expansion has not been as successful. Therefore, we need to make a significant stride forward in agriculture. All these considered, the global food supply was increased enormously in recent farming techniques pertaining to the Green Revolution. The influence on the environment and on ecosystem services was also unanticipated and harmful, which highlighted the need for more sustainable agriculture techniques [1, 2]. Excessive and improper use of fertilizers and pesticides in soil and surface waterways has been well established to increase nutrients and toxins, incur health and water purification expenses and decrease fishing and recreational alternatives. Agricultural practices that contribute to the eutrophication of aquatic habitats by degrading soil quality and may require increased fertilization, irrigation and energy expenses to maintain productivity in degraded soils. They also destroy profitable insects and wild animals.

Nanotechnology may be possibly the best solutions to eradicating the problems. Nanotechnology has got gained strenuous attention these days due to that will its diverse applications. Aside from that, there will be numerous future benefits, such as improved food quality and defense, a reduction in agricultural inputs, enrichment by absorbing nanoscale nutrients from your soil, etc. the usage of nanotechnology to be a persistent burden. For example, nanopesticides and nanofertilizers can assist in generating returns without decontaminating soils, water, and protecting against a few insect infestations and microbial issues [3]. As a result of nanotechnology, new agrochemical agents and totally new delivery methods are available to boost crop yield, and pesticide applications are expected to be reduced. Agriculture can benefit from nanotechnology in a number of ways, such as: nanoformulations of agrochemicals for the application of pesticides and fertilizers for crop progress, the use of nanosensors in crop protection with the identification of diseases, and nanodevices for genetically modifying crops and postharvest management, among other things [4, 5].

Nanotechnology has a lot of promise in agriculture, but there are still a few concerns that need to be addressed as part of the risk assessment. Nanoparticle attractants made from biopolymers such as proteins and carbohydrates have a low impact on human health as well as the environment in this regard [6]. When it comes to gardening products, nanotechnology may be used at every stage of the process, from manufacture to storage to labeling and distribution [7]. Because of its potential to enhance plant absorption of nutrients, detect disease, and manage pest infestations, nanotechnology can transform the agricultural and food sector.

2. Nanoparticles

The word "nano" is derived from the Latin nanus, which means dwarf or little. Nanoparticles (NPs) are tiny particles with a diameter of 1–100 nm. When a particle has a diameter between 1 and 100 nm, it is classified as a primary nanoparticle [8]. Nanoparticles, whether or not regarding normal or perhaps made origins have got inside the array of 1–100 nm inside one or more dimensions. A nanometer (nm) is a SI unit of length equal to 10–9 meter. NMs are materials with a length of 1–1000 nm in at least one dimension; however, they are usually characterized as having a 1–100 nm diameter. The U.S. Food and Drug Administration (USFDA) also describes NMs as "materials with an area of roughly one dimension and depending on dimensions of approximately 1 to 100 nm" [9]. To put it another way: according to International Organization for Standardization (ISO), NMs are "materials with any external or internal nanoscale dimension or surface structure at the nanoscale" [10].

Typically, nanometer will be a single billionth of the meter. Nanoemulsion, carbon dioxide nanotubes, quantum dots, nanorods, small and also nano- encapsulation and so forth. Morphology-aspect proportion or perhaps dimensions, hydrophobicity, solubility-release regarding dangerous types, surface or perhaps roughness, surface area types contaminations or perhaps adsorption, in the course of activity or perhaps historical past, reactive O2 types (ROS) O2 / H2O, ability to make ROS, construction, structure, competing holding web sites together with the receptor and also dispersal and also aggregation will be the crucial qualities regarding nanoparticles. Nanoparticles have several unique properties including a greater charge density and reactivity, considerably more strength, increased heat resistance, a lower melting point, and different permanent magnetic properties linked with nano-clusters. Distinctions inside the uncovered surface area regarding diverse nanoparticles cause differences inside atomic syndication throughout the nanoparticles, which often affect the particular electron exchange fee kinetics among metallic nanoparticles and also matching adsorbed types. These types of distinctive qualities provide the subsequent benefits to nanoparticles within farming, for example, greater solubility within the suspension, greater transmission associated with seedling jackets as well as consequently rising origins, much better bioavailability associated with substances towards the seedling radicals, supplying real focus as well as managed discharge associated with fertilizers or even pesticides within reaction to particular problems, enhanced specific exercise as well as eco-friendly along with security as well as calm transportation.

Nanoparticles tend to be seen as a distinctive bodily as well as chemical substance functions such as surface area, pore size, particle morphology, and reactivity because of their rigorous programs within the farming area. Nanoparticles are used in nano fertilizer, nano-pesticides as well as herbicides that are helpful to improve plants development, to manage extreme utilizes associated with chemical substances fertilizers as well as improve survivability towards biotic tension. The effects of various nanoparticles on plant development and phytotoxicity have been documented by a number of substances such as magnetite (Fe3O4) nanoparticles, alumina, zinc, and additionally zinc oxide in relation to seed germination [11]. Wheat can benefit from the addition of nanoparticles to their environment. The particular Zn takes on essential function inside place metabolic rate simply by influencing the actions regarding hydrogenase and also carbonic anhydrase, stabilization regarding ribosomal fractions and also activity regarding cytochrome. Place digestive enzymes stimulated simply by Zn get excited about carbs metabolic rate, servicing with the strength regarding cell filters, necessary protein activity, rules regarding auxin activity and also pollen creation. Magnesium (Mg) is involved with numerous physical as well as biochemical activities; it's an important component concerning growing development as well as improvement and performs a vital part within grow support systems within abiotic tension circumstances. The actual common perform associated with Mg within vegetation is most likely it's part since the main atom from the chlorophyll molecule within the light-absorbing complicated associated with chloroplasts and it is a factor to photosynthetic fixation associated with CO2.

3. Potential application of nanoparticles as fertilizer in agriculture

Having the limited resources, development of agriculture for the higher growth now becoming the management practices, and fertilizer management has proved the significant in a remarkable way [12, 13]. Considering the novel method of fertilizer application as a form of nanoparticles has stimulated the attributes of plant growth by the upgradation of soil system [14]. As the field condition, fertilizers variation, pH of soil etc. are being the major determinant but the slow-release mechanism of nutrients improves the efficiency of fertilizer use efficiency [15]. The activities of plant root system also accelerated by the use of nanoparticles [16]. Researchers [17] addressed chemical fertilizers as the prime factor for crop production while at the same time it also degraded the soil fertility. But the controlled use of nano-fertilizer has improved the physio-chemical attributes of the crop plants in every possible way from synthesis to metabolism. It has made more efficient practices to improve the system of agricultural practices with the new idea like precision farming involving with the technology like slow release, quick release, specific release, moisture release, heat release, pH release, ultrasound release etc. [18]. Although nanotechnology in agriculture has proved the blessings but it also has come with great risk for every living communities [19]. Therefore, to implement the advancement of agriculture sector for developing countries like Bangladesh need to grasp this huge potential of nanotechnology without delay.

4. Nanotechnology in pesticides use

To meet the hunger of the over populated countries higher yield is the prime concern and to meet the yield potential use of pesticide has covered the whole system of farming [20]. Pesticides are chemical substances enormously used to eliminate and control the harmful organisms that cause economic damage to agricultural production [21]. Every year insect pests and plant pathogens cause significant crop loss, which is around 14% and 13%, respectively, with an approximate value of U.S. $2,000 billion globally [22]. It has been reported that a minimal amount, approximately less than 0.1%, of pesticide reaches the sites of action due to loss of pesticide in the air during the application, and as run-off, spray drift, off-target deposition, and photo-degradation, the remaining bulk contaminates the surrounding environment [23, 24]. Again, these toxic chemicals are responsible for various health issues such as neurological effects, respiratory diseases, cancer, Parkinson's disorder, fetal diseases, infertility, diabetes, and genetic disorders [25]. With the rising demand for pesticides throughout the world to minimize the effects of pathogens and pests, measures should be taken to reduce pesticides' excessive application by finding appropriate alternatives. Due to the extensive use of conventional pesticides, bioaccumulation, which is caused due to biomagnifying of persistent organic pollutants and the development of resistance in the target pests, is a major concern in this generation [26, 27]. Therefore, a craze has already been started to minimize the excessive pesticide use. With the application of nanotechnology in manufacturing various nano-based pesticides, we can easily overcome these limitations [28]. Nanotechnology can reduce the application of high amounts of pesticides as nanomaterials have a high surface area with enhanced reactivity, thus lowering the cost with increasing yields [29]. Thus, the less frequent application is good for costs and human and environmental safety, ultimately a great asset for sustainable agriculture [30].

Nano based pesticides, fungicides, herbicides, molluscicides, nematicides, miticide, and nanoparticulated growth regulators are effective against various pests such as insects, rodents, weeds, fungi, viruses, bacteria, and mites [31–34]. They are also eco-friendly as they increase the formulation properties, including dispersion of water, chemical solidity, adhesion, permeability, and finally, controlled-release [35, 36]. The solubility rate of poorly soluble active ingredient can be increased with nano pesticide formulations, which ultimately helps the active ingredient release slowly and effectively. For example, nanoparticle formation like nanoencapsulation of agrochemicals such as insecticide or pesticide can increase the absorption rate and the slow and efficient release of various agrochemicals to a particular host plant for pest control [37, 38]. A small amount of nano-pesticides gives better crop protection because of their high reactivity at Nanoscale compared to their bulk counterparts [39]. For example, for the development of various nano-pesticides, we can use some effective chemical ingredients like silica, silver, carbon, and aluminium silicate [40].

Nanoparticles can be used effectively in pest control through the use of porous silica loaded with water-soluble pesticide for the ability to slow-release [41]. The cuticular lipid barrier is found in insect pests, which can be obstructed by the nano-silica component [42]. Nanoscale alumina has been used effectively against two insect pests: *Rhyzopertha dominica* (F.) and *Sitophilus oryzae* (L.), and these nano aluminium particles have shown high mortality [43]. *S. oryzae* (L.) can also be controlled with Ag NPs, synthesized from leaf extracts of *Euphorbia prostrate* with 100% mortality rate [44]. On the other hand, combinative applications of compounds like nanoparticles of silver-zinc combined were applied against one of the destructive pest, *Aphis nerii*, with a high level of mortality [45]. Ag NPs decrease the longevity of cotton bollworm (*Helicoverpa armigera*) upon treating larvae and pupae [46]. In addition, CuO NPs controls cotton leafworm larvae (*Spodoptera littorals*) with mortality of 100% [47]. The application of *Bacillus thuringiensis*-coated ZnO NPs is effective nano pesticides that delay the larval and pupal development period of cowpea weevil (*Callosobruchus maculatus*) [48]. A large number of chemical companies have started marketing nanoparticle-based pesticides in recent years, for example, Subdue M.A.X.X., Primo M.A.X.X., Banner M.A.X.X., Ospray's Chyella, and Penncap-M [49, 50]. Beyond any doubt, nanotechnology in pesticides is going to take over the conventional method of pesticide use for its efficiency. We also need to assess the credibility of nanomaterials in pesticide as it is totally new and directly related to the environment [51].

Nanotechnology is a significant research strategy which enables easy understanding of technology for the modern world. From the enormous efficiency of nanotechnology pesticide based on nanoparticles, encapsulation of nanoparticles or nanoparticle-based DNA transfer to enhance the pest resistant are some examples of smart and precision farming [52]. Because it can stimulate the improvement of farming by means of application in nanoscale strategies [53]. Which will enable the hazardous use of chemicals in farming sector by stimulating the approach of farmers to implement the precision farming. The present strategies of farming are the application of pesticide without considering the actual efficiency or persistent. That is why nanotechnology-based use of pesticides like nano-encapsulation, controlled release mechanism could be useful for the betterment of sustainable agriculture [54].

5. Use of nanobiosensors

The Advancement in the 21st century of agricultural science new ideas like nanotechnology evolved the system of cultural practices. Now a novel strategy for this aspect is the nano-biosensors. These nanoscale miniature devices are used to detect analytes at extremely low concentrations. It is a method of integrated approach with the combination of computer, electronics or nano-sciences in respect to the concern of biology [55, 56]. The sensors are being used to control moisture and pH level of soil, monitor temperature, crop nutrient status, insects, plant diseases, weeds etc., ion detection with the effectiveness of fertilizer or pesticide application strategy [57]. This real-time monitoring is accomplished by deploying wireless nanosensor networks over cultivated fields, which provide critical data for agronomic intelligence operations such as crop planting and harvesting at the appropriate times. The main strategy of this mechanism is to being the cost-effective production in agriculture by promoting the techniques of less input for the farming activities [58]. Nanobarcodes and nanoprocessing might potentially be used to track agricultural product quality. Scientists at Cornell University exploited the notion of supermarket barcodes to decode and identify diseases in a cheap, efficient, faster, and easy way. They developed minuscule probes, sometimes known as nanobarcodes, that may be used to track numerous diseases in a farm and be identified using any fluorescent-based equipment.

6. Nanomaterials for soil remediation

Soil is an inevitable medium for plant growth and food production; also, it operates planetary processes for the existence of life on earth. That is why soil is the most crucial component for the terrestrial ecosystem to flourish [59, 60]. Soil operates various vital events like biogeochemical cycles, water cycle, earth's climate, pollutant detoxification, biogenic gas regulation, ecosystem restoration, and biodiversity maintenance [61, 62]. Soil can be contaminated by various chemicals like heavy metals, pesticides, and POPs that can be remediated effectively with nanomaterials' help. For example, nano-based materials can be used to convert heavy metals to their less toxic forms, pesticide degradation, and bioremediation of contaminated soil. Besides, nano-based sensors are useful components for detecting harmful pesticide residue in the soil, like detecting Mn impurities with grapheme nanoribbon [63, 64].

For soil remediation, conventional physical and chemical methods are available, but there is a risk of secondary contamination due to these remediating agents' high quantity uses [65]. Again microbial-based soil remediation is eco-friendly but not sufficient for higher costs [66]. These limitations can be easily eliminated with nano-based techniques such as nano fertilizers, nano biosensors, nano pesticides, and different nano-remediation processes [55, 67, 68]. Conventional soil remediation methods are mainly in situ and ex situ types where nano-remediation is normally on-site method without transportation of soil, which makes this cost-effective [69, 70].

Nanoparticles possess various mechanisms such as redox reactions, adsorption, ion exchange, surface complexation, and electrostatic interaction, which are useful for the adsorption and degradation of pollutants [71]. Moreover, other features include lower temperature modification, shorter interparticle diffusion distance, and multiple surface chemistry that make these materials appropriate catalysts for the remission of the concerned soil pollutants [72]. Nanoparticles are very much fruitful for the degradation of common industrial contaminants such as chlorinated organic compounds, petroleum nano aromatics, nitrates, heavy metals (arsenic (As) lead (Pb), copper (Cu), zinc (Zn), nickel (Ni), cadmium (Cd)), insecticides, and dyes [73–76]. For instance, specific organic and inorganic compounds such as natural short-ordered aluminosilicate, the surface of titanium oxide, and humic acids can be coupled with Ni through a multiwalled carbon nanotube. These components are effective nano-bioremediation for the sustainable agricultural system [77, 78].

Nano-scale zero-valent iron (nZVI), titanium dioxide (TiO2), zinc oxide (ZnO), multiwalled carbon nanotubes (MWCNTs), fullerenes, bimetallic nanoparticles are widely used NPs for soil remediation because of their large surface area, high reactivity, and reduction capability [70, 79–81]. Surface-modified nano-scale carbon black can reduce the bioavailability of Cu and Zn; also, nanometer hydroxyapatite can remove Cd pollution from the soil, which promotes plant growth [82, 83]. On the other hand, nanometer zeolite can remove Cu and Pb; both are organic and heavy metal pollutants of soil [84]. Some researchers [85, 86] showed that Cd and Zn pollution could be repaired with the help of a ferric tetroxide nanometer.

Apart from all positive impacts, nanomaterials caused toxic effects on organisms dependent on soil [87]. For instance, copper nanoparticles negatively impact rats, as copper's toxicity is related to the particle size [88]. Again, some heavy metal ions can be dissolved with metal nanomaterials, which is toxic for the ecology. Nano-$TiO2$ and its byproducts affected the antioxidant system and oxidative stress reaction of earthworms, one of the essential soil organisms [89, 90]. For this reason, we should pay more to keep an eye on the biological toxicity of nanomaterials used in soil remediation. Though the development of the appropriate use of nanotechnology for remediation of polluted soils is essential with the help of numerous uses of nanomaterials, we also need a comprehensive understanding of the human and environmental risk–benefit balance by using these nanomaterials [91].

7. Fate of nanomaterials in soil

In assumption, soil is meant to be the biggest receiver of NPs. When nanoparticles or nanoformulations are given to plants, the substance eventually makes its way to the soil, where it may be used. Because of its proximity to plant roots and microbes, the nanomaterial establishes a unique connection with them once it is in the soil. In addition to soil, which is the most abundant source of natural nanoparticles, both as primary particles and aggregates, it is also regarded an externally significant environmental matrix. Dissolution, transformation, and aggregation/disaggregation are among of the mechanisms that regulate the fate of NMs in soils. In soils, several of the mechanisms that determine NM fate and behavior, such as straining, deposition/mobilization and diffusive transport are substantially different (**Figure 1**). The significance of these factors varies depending on the NM and soil conditions [92, 93]. Because dissolution destroys them and aligns their destiny and bioavailability with the soluble components, dissolution may be critical for some NMs. Examples include the fast dissolving of ZnO, which is likely to be ephemeral in soil unless coated with compounds that prevent dissolution. When it comes to dissolving, pH is the most essential factor to take into consideration.

For example, researchers [92] used a worldwide database to compare soil saturation extract pH and ionic strength to NMs critical coagulation concentrations. Since the pH and Ionic Strength of most soil solutions are below the critical coagulation concentration of most nanomaterials, homoaggregation would be sluggish in most soils. As in aquatic contexts, heteroaggregation is anticipated to play a major role in soils because soil porewaters frequently include larger quantities of natural colloids in suspension. In most soils, NM condensation will occur in the topsoil with limited transport to the depths, resulting in increased straining (**Figure 1**). In soil porewaters, NOM has been reported to stabilize NMs and prevent both homo- and heteroaggregation, according to several studies [94]. NM movement in soils has been studied in a limited number of researches, and this is likely the most important gap in understanding the true danger of NM transport. NM transport researches in soils have evolved from utilizing inert stationary phases (e.g., quartz beads) in columns to employing real soils in the recent decade. It appears that the CNTs are retained in soils due to their large aspect ratio, resulting in considerable straining. There is a high concentration of fullerenes in soils due to interactions with soil organic matter.

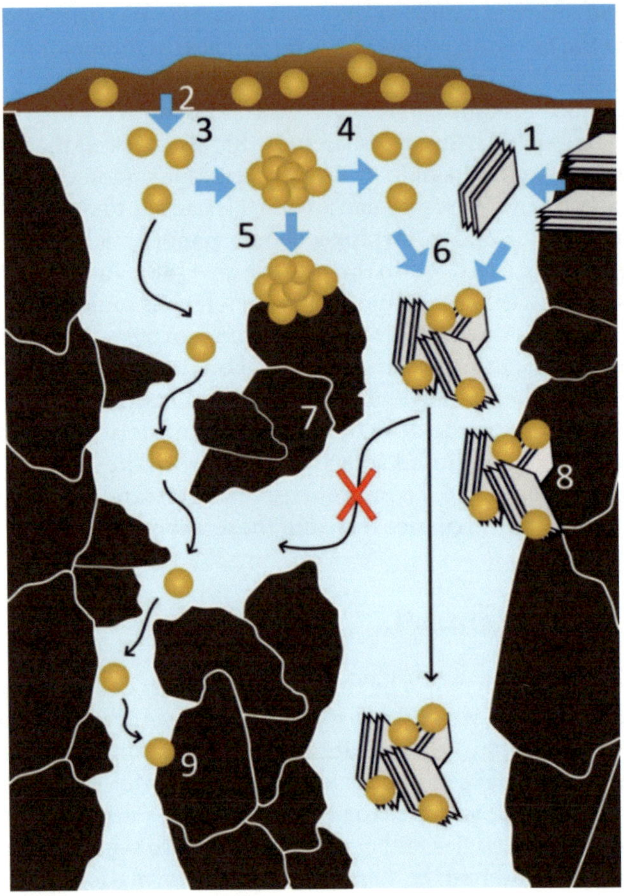

Figure 1.
The key mechanisms that determine the fate of natural colloids and nanomaterials in soils. 1, colloid generation; 2, engineered nanomaterial leaching from biosolids; 3, homoaggregation; 4, fragmentation; 5, sedimentation; 6, heteroaggregation; 7, size exclusion; 8, straining; 9, deposition; 10, convective transport [92].

8. Effect of nanomaterials on soil/plant systems

Nanotechnology has opened up new avenues for increasing nutrient efficiency and lowering environmental protection expenditures. The fact that fertilizer usage efficiency is just 20–50% for nitrogen and 10–25% for phosphorus is alarming [95]. According to researchers [96], the introduction of nanofertilizers as an alternative to traditional fertilizers would remove nutrient buildup in soils, therefore eliminating eutrophication and drinking water pollution. The main idea is to improve the efficiency of native and applied phosphorus in soils to keep the ratio of applied and plant absorption P near unity. Because nearly all P fertilizers contain heavy metals and, more significantly, deliver P to plants in accessible forms, there is a need to regulate critical and harmful components linked with phosphorous in the pedosphere–hydrosphere continuum. Nanofertilizers increased the quality of agricultural goods, eliminated environmental risks, and needed less fertilizer than conventional fertilizers [97]. Because the rate of release of absorbed nitrogen (or fertilizers compounds) is much slower than that of adsorbed ionic forms of nitrogen, zeolites could be used for nitrogen capture and storage [98]. Researchers [99] noticed that zeolite chips containing urea in their cavities can be used as a slow-release nitrogen fertilizer material. When zeolites were loaded with nitrogen, potassium, phosphorous, calcium, and a set of minor and trace nutrients, few researchers [100] discovered that the honeycomb-like layered crystal network slowly released nutritional ions "on demand."

According to another researcher [101], application of a nanocomposite comprising N, P, K, certain micronutrients, mannose, and amino acids improved nutrient absorption in grain crops. A group of researchers [41] used zinc–aluminum layered double-hydroxide nanomaterials with plant growth regulators and discovered that the products released chemicals in a regulated way. These studies showed that nanotechnology may be used to build advanced supply tools with great success. There are carbonaceous chemicals that are secreted into the soil that allow N and/or P mineralization from organic matter and P mineralization from soil inorganic colloids in nutrient-depleted soils. As environmental signals, these root exudates may be utilized to make nanobiosensors, which could then be integrated into new nanofertilizers, according to the authors' hypothesis.

It's well-known that current fertilizers create soil acidity, alter soil carbon profiles, harm beneficial microorganisms, weather clay minerals, and collect heavy metals in the soil. As receptacles, novel nanofertilizers use plant-nutrient ions intercalating or adsorbing on clay minerals. Salts make up the majority of current fertilizers, with one component consisting of plant-nutrient ion(s), whereas the other component isn't particularly beneficial or harmful at all. To enhance soil structure, reduce salt concentration, and promote crop development in salt-affected soils, nanotechnology can be utilized to improve soil structure. The following are some areas where research might be initiated: CaCO3 solubilization, Na2CO3 prevention, adding K+ to clay minerals, and increasing precipitation are all examples of ways to reduce salt concentration in soil solution, improve drainage, and/or replace Na + with Ca2+ and/or K+.

In order to determine the influence of nanoparticles on soil microbial activity, soil respiration and enzymatic activity must be measured. Soil enzymatic activity and bacterial abundance may be affected by metallic nanoparticles [102]. They can also cause free radical damage to bacteria' cell membranes, DNA and mitochondria. Even beneficial microorganism communities may be threatened by the introduction of nanoparticles (NPs) into the natural environment. In flooded paddy soil, TiO2 and CuO nanoparticles reduced soil microbial biomass and enzymatic activity, as well as their community structure. Increased Fe3O4 nanoparticle concentration dramatically reduced the number of bacteria in soil, and produced cavities, holes and membrane breakdown in the microorganisms [103].

Bipolaris sorokiniana and *Magnaporthe grisea* were exposed to silver ions and silver nanoparticles to determine their effects [104]. These treatments effectively inhibited colonization of both fungi, with an EC50 much lower than the ionic Ag treatments. Scientists have demonstrated antibacterial activity of Ag nanoparticles and polyvinylpyrrolidone (PVP) against three types of bacteria [105]. Researchers have shown that Zinc oxide nanoparticles (ZnO NP) are as antibacterial as silver nanoparticles (AgNPs). Sulfur dioxide (ZnO) was typically more toxic to bacteria in the Gram-positive group than the Gram-negative group. *Staphylococcus aureus* was treated for 8 hours, and *Salmonella typhimurium* for 4 hours [106]. When it came to *Botrytis cinerea* and *Penicillium expansum* colonization, the S NPs (35 nm) were shown to be more efficient than the larger particles.

9. Nanomaterials to mitigate environmental stresses in plants

Plants are sessile organisms and undergo abiotic stressors that impact their development and production throughout their life cycle. In response to environmental stressors, plants generate defensive mechanisms at multiple levels through modification of their biochemical and morphological routes as well as their molecular pathways (the changing of genetic expressions). But these are not sufficient to annul all the adverse effects of environmental stress. The salinity reduces, for example, the osmotic potential of the soil, resulting in food disequilibrium. Improve ionic toxicity negatively impacts many important biochemical or physiological activities including photosynthesis, protein synthesis and lipid metabolism. The rising world population and the concomitant decline in food supply, with ongoing environmental changes, are currently in a difficult state. Therefore, scientists' main focus is to develop strategies to expedite the plant adaptation to environmental changes.

In the worldwide scenario, salt stress alone reduces crop yield by roughly 23 per cent according to current agricultural practices. In previous study on nano-SiO2 treatment on tomatoes and squash plants, there have been numerous beneficial results on the usage of nano fertiliser in salty circumstances [107, 108]. The use of silica nanoparticles increases plant tolerance to drought stress by promoting plants' agronomic parameters, physiology, biochemistry, delay senescence, and maintained water status of plants exposed to the water-deficit condition [109, 110]. *Crataegus sp*. has enhanced dryness tolerance with varied concentrations of silica nanoparticles, changing their physiological and biochemical processes [111]. Researchers think that growing agricultural plants with shorter life cycles is particularly efficient in areas susceptible to drought or flash-flood here early crop maturation is a critical component for sustained crop output [112]. Studies revealed that the life cycle of the wheat crop used in nano fertilizers is considerably shorter than the traditional wheat crop used in fertilizers, which is 130 days compared with 170 days (date of sowing to yield production) [113].

Although several investigations of the use of nanomaterials to plant development have been carried out during stress environments, the fundamental components remain mostly unexplored. However, researchers believe that, under unfavorable environmental circumstances, the impacts of nanomaterials on crop development are partially attributable to the enhanced enzyme activity. The activity of antioxidant enzymes such as superoxide dismutase (SOD), catalase (CAT), and peroxidase (POD) are regulated by nanoparticles [114]. Increased SOD activities have been observed by the application of TiO2 nanoparticles on onion seedlings [115]. At lower levels of TiO2 the activity of the enzyme is greater. The buildup of free proline and amino acids is escalating by nanomaterials (nano-SiO2 and nano-ZnO). The consumption of nutrients and water might also rise. The use of these nanoparticles further enhances the activity of the antioxidant enzymes such as SOD, CAT, POD and reductase nitrates. Nanomaterials can also control the expression of stress genes. For example, microscope research showed that silver nanoparticles in *Arabidopsis* can regulate a number of genetic expressions [116]. These genetic responses, which are produced by nanomaterials, are therefore directly related to plant stress defense.

10. Nanomaterials in plant defense mechanisms

Plants represent the boundary between the environment and the biosphere, thus understanding how nanomaterials influence them is crucial for ecological evaluations and assessments of environmental impact. Terrestrial plants can be threatened by metal-based nanoparticles (NPs); yet, little is known about plant defense systems that could combat nanotoxicity. When cells are subjected to nanoparticles and oxidative pressure develops, the equilibrium between cell function as well as antioxidative defense systems is altered. A group of researchers [3] described

Nanoparticles	Mode of action	References
Chitosan	Total phenolics and NO signaling molecules are elevated, as is the expression of defense-related genes, such as numerous antioxidant enzymes.	[118]
TiO_2	SOD, catalase, and peroxidase (POD) activities are increased, and reactive oxygen free radical buildup is reduced.	[119]
Multi-walled carbon nanotubes (MWCNTs)	Antimicrobial pathogens cannot complete their life cycle because ROS defense response cascade is activated. ROS, such as super peroxides and H_2O_2, are generated.	[120, 121]
Silicon	An improved resistance against fungus in maize	[122]
Copper oxide	Raise the levels of SOD and CAT, as well as lipid peroxidase	[123]
Cerium oxide	Rice seedlings are exposed to lipid peroxidation and photosynthetic stress when exposed to CeO_2 (modifications of antioxidant defense system)	[124]
Zinc oxide	GSH levels and CAT activity are both higher.	[125]

Table 1.
Nanoparticles and its mode of action.

cell membrane damage due to oxidative stress, as well as DNA degradation are all caused by biochemical factors that produce unnecessarily high reactive oxygen species (ROS). Different defensive mechanisms can be triggered by plants in response to stress [117]. As an example, using nanoparticles to boost plant defenses is one of the most intriguing aspects of this technology (**Table 1**). An enzyme and a non-enzymatic agent are used in plants' antioxidant defense system. These agents include SOD, CAT, APX, ascorbate peroxidase (APX), and glutathione reductase (GRT) (GR).

It has been demonstrated that nanoparticles of cerium oxide imitate enzymes for scavenging. This feature increases the plant's defensive system. As a result, microbial pathogens are prevented from completing their life cycle by multiwalled carbon nanotubes (MWCNTs). Changes in enzymes are prevalent as a result of fluctuations in ROS levels [126]. ROS play a major role in the start of plant disease resistance responses, since they are essential signals for resolving defensive gene installation. To further understand plant defense mechanisms against nanoparticles, more research is needed.

11. Nanotechnology in food industry

Food production must double by 2050 to satisfy the demands of the world's increasing population, food production must double by 2050, and new strategies are needed to fight hunger [127]. The rising global human population has resulted in a larger population to feed, and agricultural production has not kept pace with this growth. This imbalance has shown the actual need for food preservation for food items to reach people worldwide. The establishment of nanotechnology in the food sector has made it easier for food to be transported to various areas globally by increasing most food items' shelf life. The latest developments in nanostructured materials that significantly affect the food sector are novel methods in food nanotechnology (**Table 2**). Nanotechnology in today's food sector has played a significant role in food processing, food packaging and food preservation. Many areas of food science have been revolutionized by the fast growth of nanotechnology,

Nano-Technology	Description	Example	Reference
Nanostructures food ingredients	Nano-engineered additives	Titanium dioxide (TiO2) is a Nano-engineered additive that is used as a colorant, and an antibacterial agent.	[128]
Nano-Emulsion	Nano structured Emulsion	The Nano-Emulsion method is used to produce low-fat mayonnaise and ice cream that are as creamy as their full-fat counterparts, providing customers with healthier choices.	[129]
Nano-encapsulation	Delivery systems for additives and bioactive	Nano-encapsulation is the process by which bioactive chemicals are incorporated, absorbed, or dispersed.	[130, 131]
Food packaging	Silver, Zinc oxide, Titanium oxide, Silver oxide used in the food packaging.	In green asparagus, silver nanoparticles inhibit the development of aerobic psychrotrophic bacteria.	[132]
Nanosensors and nanobiosensors	Quality control and food safety assist in the detection of any subtle changes in food color as well as any gasses produced due to spoilage	Gold, platinum, and palladium are used to make the gas sensors. The gold-based nanoparticles can detect the aflatoxin B1 toxin present in milk.	[133]
Food processing	Fortification	Incorporation of nutraceuticals, viscosifying and evaporating agents, vitamin and mineral fortification	[134]

Table 2.
Current uses of nanotechnology in food industry.

particularly those involving food processing, packaging, storage, transportation, functioning, and other safety concerns. A wide range of nanostructured materials (NSMs), from inorganic metal, metal oxides, and their nanocomposites to nano-organic materials with bioactive agents, has been applied to the food industry. **Figure 2** shows the application of nanotechnology in the food business [136].

12. Nanomaterials for recycling agricultural waste

Demand for agricultural goods is rising rapidly as the population grows. More food items are being produced to satisfy this increasing demand, resulting in a rise in waste materials. Waste is a significant issue throughout the globe, and it is produced by a variety of agricultural, industrial, and urban activities. Agricultural wastes are such kinds of wastes derived from various agricultural activities, including processing raw agricultural products; plant debris; excessive use of pesticides and fertilizers that enter into our ecosystem; wastes from animal farms and slaughterhouses; salt and silt drained from fields and finally harvest wastes. In other words, these are leftovers from the production and processing of raw agricultural goods, including fruits, vegetables, meat, poultry, dairy products, and crops [137]. Large amount-of agricultural wastes are generated every year throughout the world that can be solid, liquid, or slurries in form depending on the agricultural activities, posing a threat to the environment (**Table 3**). We are exploiting our environment using excess amounts of agrochemicals like pesticides and fertilizers every year.

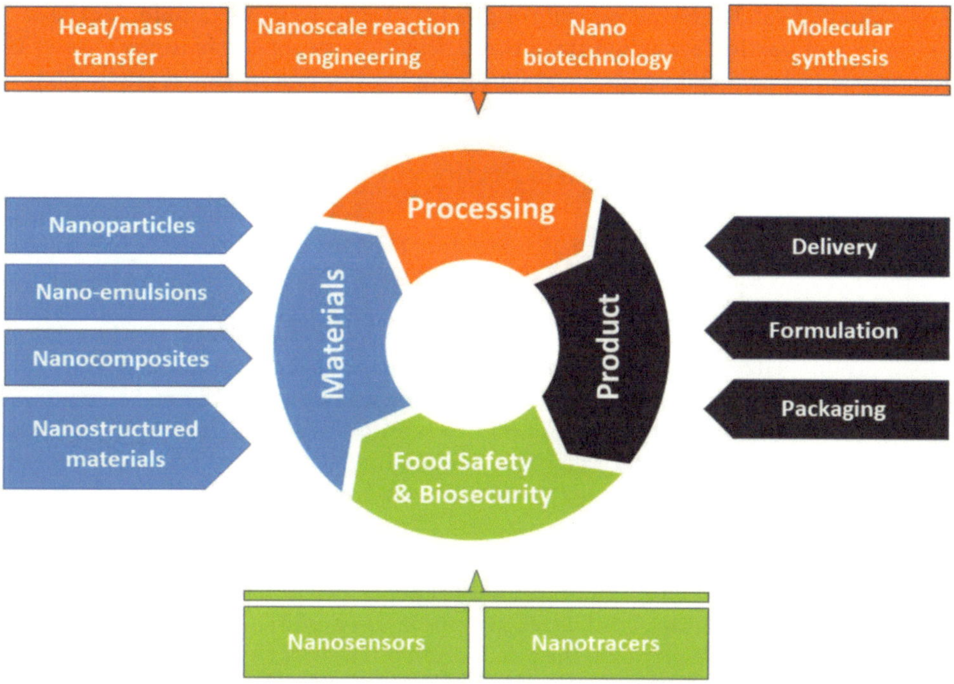

Figure 2.
Application areas of nanotechnology in food industry [135].

Classification	Origin point	Wastages
Garbage	Domiciliary/merchandise	Kitchen waste (food), poultry/slaughter house
Settlings	Domiciliary/retailer	Dead leaves, wood wool, shredders waste, waste tyre rubber, polyethelene, plastic bottles, etc.
Ashes	Households/industries	Rice husk ashes, coal fly ashes
Pharmaceuticals	Industries/households	Expired medicines, scraps etc.
Waste microbes	Industries/households	Metal ion-treated cells, fungi refusals, bacterial cellulose discard, etc.
Agricultural waste	Beverage industries/food processing industries	Palm oil mill effluent, defatted seeds, orange peels, etc.

Table 3.
Classification of waste as per their source of origin [138].

These chemicals are generally persistent and have a significant impact on the environment as well as human health due to the bioaccumulation in food. For sustainable agriculture, we badly need an efficient way to properly use agricultural inputs and reduce these wastes to minimize environmental pollution. Through nanotechnology, pesticides and fertilizers can be converted and reused. Some nanomaterials for the remediation of soil polluted by agrochemicals are encapsulated and slow released fertilizers and pesticides under specific conditions; controlled release of plant growth hormones and concentration of ammonium nanoparticles that can be recycled as fertilizer [139, 140].

Photocatalysis applications, coupled with nanotechnology, offer effective results and enormous possibilities for the reduction of certain harmful chemicals from various herbicides, bactericides, and fungicides (**Table 4**). For example, for the elimination of pesticide residues from water, therefore decontamination of water is effective with the process of photocatalysis coupled to a nanomaterial [153]. On the other hand, nano-sensors can detect various chemicals and toxic pollutants that are harmful to humans. The application of nanomaterials coupled to specific antibodies can generate lights that can be used to identify and quantify agrochemicals like pesticides and fertilizers [139].

Apart from this, rice husk, a by-product from rice-mill, can be an excellent source for nano-silica production, making glass materials and concrete. This renewable nano-silica ultimately reduces the rice husk disposal problem through nanomaterials. Waste from the cotton industry, such as cellulose or other low valued products like yarns and cotton balls, can be reduced with nanomaterials' help. For example, with the use of electrospinning and newly developed solvents, 100 nm-sized fibers can be produced and use as an absorbent of various fertilizers or pesticides, which is useful for targeted application at the desired time as well as location [154, 155]. Nanocellulose can be extracted from the residues of banana cultivation like pseudostem, foliar parts, and shells, which will be the replacement of certain synthetic fibers. On the other hand, gold nanoparticles, which are numerously used in semiconductors and bio-medical areas, can be synthesized from agricultural wastes of grape seed, skin, and stalks [140, 156].

From the last couple of years, the production of ethanol from maize feedstock has increased the global price and demand of maize, and researchers are working on various nano-engineered enzymes that authorize simple and cost-effective modification of cellulose into ethanol from waste plant parts [131]. Nanomaterials also inspire the metabolism of microorganisms like the efficacy of lipid extraction can be improved with the help of nanotechnology without disrupting the microalgae. Nanomaterials like calcium and magnesium oxide nanoparticles can be used successfully as biocatalyst transporters for the transesterification of oil to bio-diesel [157].

Due to the mass production of agricultural goods, many wastes are generated every year from this sector, and with the application of nanomaterials,

Agricultural wastes	Nanomaterial associated	References
Rice husk	Calcium hydroxyapetite NPs	[141]
Olive mill wastewater	Titanium (IV) oxide anatase, iron (III) oxide nanorods NPs	[142]
Waste cooking oil	Magnetic NPs, Mesoporous silica/ superparamagnetic iron oxide core shell NPs, Molybdenum oxide/Zirconia NPs	[143–145]
Water with high phosphate content	Iron NPs	[146]
Peels of Pomegranate	Silver nanoparticles	[147]
Coconut shells	Silver nanoparticles	[148]
guava	Silver nanoparticles	[149]
Sugarcane waste	Silver and Gold nanoparticles	[150]
Banana peel	Gold nanoparticles	[151]
Grapes waste	Gold nanoparticles	[152]
peanut skin extract	Iron nanoparticles	[79]

Table 4.
Nanomaterial-associated waste management.

these wastes can be reduced, reused, and recycled effectively. Also, this new technique can be an asset for poor nations having poor sanitation, water scarcity, and inadequate resources [2, 158]. When crops are harvested, additional connected problems exist, such as crop waste, nearly 80% of the farm's biomass. The production of agricultural waste is hundreds of millions of tons annually [159]. Every year, a large amount of food and agricultural goods are wasted as agro-waste throughout the globe. It estimates that about one-third of the world's food produced for human use is lost or destroyed each year [160]. Minimizing agricultural product losses reduces resource pressures and therefore reduces the need for chemical fertilizers and pesticides [161]. It is thus time to manage waste strategically to recycle, recycle and reuse agro-purpose.

Nanotechnology is now confined to the energy, food hygiene, telecommunications, agriculture, and healthcare sectors and has now covered environmental protection and waste management. Green nanoparticles production is becoming more popular in a straightforward, ecologically friendly manner. The continual deposition of agricultural wastes or byproducts in nature has become a significant concern. Nanotechnology has the potential to be used in the reduction of waste generated during agricultural production. Agricultural wastes, including natural and non-natural wastes, may also be effectively used to produce nanoparticles.

13. Conclusion and future perspectives

Nanotechnology has a wonderful possibility in agriculture. Research on nanotechnology uses in agriculture is less than ten years old. However, given the growing inadequacy of traditional farming techniques and the excess capacity of the terrestrial ecosystem demands, we have little alternative but to investigate the nanotechnologies in all agricultural sectors. New technology is generally acknowledged as essential to the creation of national prosperity.

There's been a substantial improvement upon nanoparticles dependent programs in agriculture industries. Scanty reviews can be found about the suitable utilization as well as improvement associated with eco-friendly nanoparticles in several fields. Therefore, execution associated with nanomaterials may uplift the actual farming requirements and supply advantages in various methods. However, among the main constrict may be the toxicity associated with nanoparticles. Therefore, to conquer the actual poisonous results, various logical methods are now being created. One particular technique entails the utilization of (i) Natural organizations or even their items concerning manufacturing associated with nanoparticles that type among the eco-friendly procedures about functionality associated with nanoparticles. (ii) Bioconjugation as well as encapsulation associated with nanoparticles along with bioactive substances is guaranteeing area that prevents toxicity. (iii) Nanotechnology also offers options about degrading continual chemical substances into safe as well as occasionally helpful elements. (iv) Nanotechnology may effort to supply as well as essentially improve the actual systems presently utilized in environment recognition, realizing as well as remediation. (v) To be able to obtain prosperous utilization as well as commercialization associated with nanomaterials, various knowledge ought to work with others to style biomimetic nanomaterials as well as their assessment within the agriculture field.

In conjunction with information on the agriculture production system, nanotechnology demands a solid understanding of science as well as of production and material technologies. The severity of this task can draw talented brains into a career for agriculture. To succeed in this sector, human resources require advanced training, which is urgently necessary for new instruction programs, particularly at the graduation level.

Conflict of interest

All authors wish to confirm that there is no potential conflict of interest.

Author details

Mohammad Monirul Hasan Tipu[1*], Artho Baroi[2], Juwel Rana[3], Shariful Islam[4], Raunak Jahan[5], Md. Shipon Miah[6] and Md. Asaduzzaman[7*]

1 Plant Pathology Division, Bangladesh Agricultural Research Institute (BARI), Gazipur, Bangladesh

2 Faculty of Agriculture, Bangladesh Agricultural University, Mymensingh, Bangladesh

3 Department of Nutrition and Food Engineering, Daffodil International University, Dhaka, Bangladesh

4 Department of Farm Power and Machinery, Bangladesh Agricultural University, Mymensingh, Bangladesh

5 Department of Aquaculture, Bangladesh Agricultural University (BAU), Mymensingh, Bangladesh

6 Ministry of Agriculture, Department of Agricultural Extension, Moulvibazar, Bangladesh

7 Olericulture Division, Horticulture Research Center, Bangladesh Agricultural Research Institute, Gazipur, Bangladesh

*Address all correspondence to: tipubari2013@yahoo.com and asadcbt@yahoo.com

References

[1] Tillman D, Cassman KG, Matson PA, Naylor R, Polasky S. Agricultural sustainability and intensive production practices. Nature. 2002;418:671-677. DOI: 10.1038/nature01014

[2] Mukhopadhyay SS. Nanotechnology in agriculture: prospects and constraints. Nanotechnology Science and Applications. 2014;7:63-71. DOI: 10.2147/NSA.S39409

[3] Prasad R, Bhattacharyya A, Nguyen QD. Nanotechnology in Sustainable Agriculture: Recent Developments, Challenges, and Perspectives. Front Microbiol. 2017;8:1014. DOI: 10.3389/fmicb.2017.01014

[4] Singh H, Sharma A, Bhardwaj SK, Arya SK, Bhardwaj N, Khatri M. Recent advances in the applications of nano-agrochemicals for sustainable agricultural development. Environ Sci Process Impacts. 2021;23(2):213-39. DOI: 10.1039/D0EM00404A

[5] Elmer W, White JC. The Future of Nanotechnology in Plant Pathology. Annu Rev Phytopathol. 2018;56(1): 111-33. DOI: 10.1146/annurev-phyto-080417-050108

[6] Verma ML, Dhanya BS, Sukriti, Rani V, Thakur M, Jeslin J, et al. Carbohydrate and protein based biopolymeric nanoparticles: Current status and biotechnological applications. Int J Biol Macromol. 2020;154:390-412. DOI: 10.1016/j.ijbiomac.2020.03.105

[7] Ghidan AY. Applications of Nanotechnology in Agriculture. In: Zlatev TMAAE-MSE-R, editor. Rijeka: IntechOpen. 2020. p. Ch. 4. DOI: 10.5772/intechopen.88390

[8] Potočnik J. Commission recommendation of 18 October 2011 on the definition of nanomaterial (2011/696/EU). Off J Eur Union. 2011;275:38-40

[9] FDA Guidance on Nanotechnology DOCUMENT: Guidance for Industry Considering Whether an FDA-Regulated Product Involves the Application of Nanotechnology. 2011;30(5):571-572. DOI: 10.1089/blr.2011.9787

[10] ISO/TS 80004-1:2010, Nanotechnology – Vocabulary – Part 1: Core Terms. Geneva, Switzerland: International Organization for Standardization [Internet]. 2010. Available from: https://www.iso.org/obp/ui/#iso:std:iso:tr:18401:ed-1:v1:en [Accessed: 2021-09-17]

[11] Kornarzyński K, Sujak A, Czernel G, Wiącek D. Effect of Fe3O4 nanoparticles on germination of seeds and concentration of elements in *Helianthus annuus* L. under constant magnetic field. Scientific Reports. 2020;10(1):8068. DOI: 10.1038/s41598-020-64849-w

[12] Taheri RH, Miah MS, Rabbani MG, Rahim MA. Effect of Different Application Methods of Zinc and Boron on Growth and Yield of Cabbage. European Journal of Agriculture and Food Sciences. 2020;2(4):1-4. DOI: 10.24018/ejfood.2020.2.4.96

[13] Miah MS, Taheri RH, Rabbani MG, Karim MR. Effects of Different Application Methods of Zinc and Boron on Growth and Yield of Onion. International Journal Biosciences. 2020;17(4):126-133. DOI: 10.12692/ijb/17.4.126-133

[14] Mani PK, Mondal S. Agri-nanotechniques for Plant Availability of Nutrients. In: Kole C, Kumar D, Khodakovskaya M. (eds) Plant Nanotechnology. Springer, Cham; 2016. p. 263-303. DOI: 10.1007/978-3-319-42154-4_11

[15] Shukla PK, Misra P, Kole C. Uptake, Translocation, Accumulation, Transformation, and Generational Transmission of Nanoparticles in Plants. In: Kole C, Kumar D, Khodakovskaya M. (eds) Plant Nanotechnology. Springer, Cham; 2016. p. 183-218. DOI: 10.1007/978-3-319-42154-4_8

[16] Nair R. Effects of Nanoparticles on Plant Growth and Development. In: Kole C, Kumar D, Khodakovskaya M. (eds) Plant Nanotechnology. Springer, Cham; 2016. 95-118. DOI: 10.1007/978-3-319-42154-4_5

[17] Solanki P, Bhargava A, Chhipa H, Jain N, Panwar J. Nano-fertilizers and their smart delivery system. In: Rai M, Ribeiro C, Mattoso L, Duran N (eds) Nanotechnologies in food and agriculture. Springer International Publishing, Cham. 2015; pp 81-101. DOI: 10.1007/978-3-319-14024-7_4

[18] Manjunatha SB, Biradar DP, Aladakatti YR. Nanotechnology and its application in agriculture: A review. Journal of Farm Science. 2016;29(1): 01-13

[19] Achari GA, Kowshik M. Recent Developments on Nanotechnology in Agriculture: Plant Mineral Nutrition, Health, and Interactions with Soil Microflora. Journal of Agricultural Food Chemistry. 2018;66(33):8647-8661. DOI: 10.1021/acs.jafc.8b00691

[20] Verger, Philippe JP, Boobis AR. Reevaluate Pesticides for Food Security. Science. 2013;341:717-718. DOI: 10.1126/science.1241572

[21] Rakibuzzaman M, Rahul SK, Ifaz MI, Gani O, Uddin AFMJ. Nano Technology in Agriculture: Future Aspects in Bangladesh. International Journal of Business, Social and Scientific Research. 2018;7(1):06-09

[22] Pimentel D. Pesticides and Pest Control. In: Peshin R, Dhawan A (eds.) Integrated Pest Management: Innovation-Development Process. Springer Netherlands. 2009. p. 83-87. DOI: 10.1007/978-1-4020-8992-3_3

[23] Pimentel D. Amounts of pesticides reaching target pests: Environmental impacts and ethics. J Agric Environ Ethics. 1995;8:17-29. DOI: 10.1007/BF02286399

[24] Carriger JF, Rand GM, Gardinali PR, Perry WB, Tompkins MS, Fernandez AM. Pesticides of potential ecological concern in sediment from South Florida canals: an ecological risk prioritization for aquatic arthropods. Soil and Sediment Contamination. 2006;15:21-45. DOI: 10.1080/15320380500363095

[25] Hu R, Huan X, Huang J, Li Y, Zhang C, Yin Y, Chen Z, Jin Y, Cai J, Cui F. Long and short term health effect of pesticides exposure: A cohort Study from China. PLoS One. 2015;10(6): e0128766. DOI: 10.1371/journal.pone.0128766

[26] Chareonviriyaphap T, Bangs MJ, Suwonkerd W, Kongmee M, Corbel V, Klan NR. Review of insecticide resistance and behavioural avoidance of vectors of human diseases in Thailand. Parasitol Vectors. 2013;6:280. DOI: 10.1186/1756-3305-6-280

[27] Soko W, Chimbari MJ, Mukaratirwa S. Insecticide resistance in malaria-transmitting mosquitoes in Zimbabwe: a review. Infect Dis Poverty. 2015;4:46. DOI: 10.1186/s40249-015-0076-7

[28] Chhipa H. Nanofertilizers and nanopesticides for agriculture. Environ Chem Lett. 2017;15(1):15-22. DOI: 10.1007/s10311-016-0600-4

[29] Gogos A, Knauer K, Bucheli TD. "Nanomaterials in plant protection and fertilization: current state, foreseen applications, and research priorities",

Journal of Agricultural Food Chemistry. 2012;60(39):9781-9792. DOI: 10.1021/jf302154y

[30] Prasad R, Kumar V, Prasad KS. Nanotechnology in sustainable agriculture: present concerns and future aspects. African Journal of Biotechnology. 2014;13(6):705-713. DOI: 10.5897/AJBX2013.13554

[31] Guang XY, Wang JJ, He ZG, Chen GX, Ding L, Dai JJ, Yang XH. Molluscicidal effects of nano-silver biological molluscicide and niclosamide. Zhongguo Xue Xi Chong Bing Fang Zhi Za Zhi. 2013;25(5):503-505

[32] Oliveira HC, Moreira SR, Martinez CBR, Grillo R, de Jesus MB, Fraceto LF. Nanoencapsulation enhances the post-emergence herbicidal activity of atrazine against mustard plants. PLoS One. 2015;10(7):e0132971. DOI: 10.1371/journal.pone.0132971

[33] Pereira AES, Sandoval-Herrera IE, Zavala-Betancourt SA, Oliveira HC, Ledezma-Pérez AS, Romero J, Fraceto LF. γ-Polyglutamic acid/chitosan nanoparticles for the plant growth regulator gibberellic acid: characterization and evaluation of biological activity. Carbohydr Polym. 2017;157:1862-1873. DOI: 10.1016/j.carbpol.2016.11.073

[34] Antonoglou O, Moustaka J, Adamakis ID, Sperdouli I, Pantazaki AA, Moustakas M, Samara DC. Nanobrass CuZn nanoparticles as foliar spray nonphytotoxic fungicides. ACS Appl Mater Interfaces. 2018;10(5):4450-4461. DOI: 10.1021/acsami.7b17017

[35] Zhao X, Cui H, Wang Y, Sun C, Cui B, Zeng Z. Development strategies and prospects of nano-based smart pesticide formulation. Journal of Agricultural Food Chemistry. 2017;66(26):6504-6512. DOI: 10.1021/acs.jafc.7b02004

[36] Prasad R, Kumar V, Kumar M, Choudhary D, editors. Nanobiotechnology in Bioformulations. Springer International Publishing. 2019. DOI: 10.1007/978-3-030-17061-5

[37] Scrinis G, Lyons K. The emerging nano-corporate paradigm: nanotechnology and the transformation of nature, food and agri-food systems. International Journal of Sociology of Agriculture and Food. 2007;15:22-44. DOI: 10.48416/ijsaf.v15i2.293

[38] Kah M, Beulke S, Tiede K, Hofmann T. Nanopesticides: State of knowledge, environmental fate and exposure modelling. Critical Reviews in Environmental Science and Technology. 2012;43:1823-1867. DOI: 10.1080/10643389.2012.671750

[39] Debnath N, Das S, Seth D, Chandra R, Bhattacharya S, Goswami A. Entomotoxic effect of silica nanoparticles against *Sitophilus oryzae* (L.). J Pest Science. 2011;84:99-105. DOI: 10.1007/s10340-010-0332-3

[40] Singh NB, Amist N, Yadav K, Singh D, Pandey JK, Singh SC. Zinc oxide nanoparticles as fertilizer for the germination, growth and metabolism of vegetable crops. Journal of Nanoengineering and Nano manufacturing. 2013;3:353-364. DOI: 10.1166/jnan.2013.1156

[41] Liu X, Feng Z, Zhang S, Zhang J, Xiao Q, Wang Y. Preparation and testing of cementing nano-subnano composites of slower controlled release of fertilizers. Scientia Agricultura Sinica. 2006;39:1598-1604. DOI: 10.1016/S1671-2927(06)60113-2

[42] Ulrichs C, Mewis I, Goswami A. Crop diversification aiming nutritional security in West Bengal - Biotechnology of stinging capsules in nature's waterblooms. Ann Tech Issue of State Agri Technologists Service Assoc. 2005. p. 1-18

[43] Stadler T, Buteler M, Weaver DK. Novel use of nanostructured alumina as an insecticide. Pest Management Science. 2010;66:577-579. DOI: 10.1002/ps.1915

[44] Zahir AA, Bagavan A, Kamaraj C, Elango G, Rahuman AA. Efficacy of plant-mediated synthesized silver nanoparticles against *Sitophilus oryzae*. J Biopest. 2012;5:95-102

[45] Rouhani M, Sami MA, Kalantari S. Insecticide effect of silver and zinc nanoparticles against *Aphis nerii* Boyer De Fonscolombe (Hemiptera: Aphididae). Chilean journal of agricultural research. 2012;72:590-594. DOI: 10.4067/S0718-583920120 00400020

[46] Devi DG, Murugan K, Selvam PC. Green synthesis of silver nanoparticles using *Euphorbia hirta* (Euphorbiaceae) leaf extract against crop pest of cotton bollworm, Helicoverpa armigera (Lepidoptera: Noctuidae). J Biopest. 2014;7:54-66.

[47] Shaker AM, Zaki AH, Rahim AEF, Khedr MH. Novel CuO nanoparticles for pest man agement and pesticides photo-degradation. Adv Environ Biol. 2016;10:274-283

[48] Malaikozhundan B, Vaseeharan B, Vijayakumar S, Thangaraj MP. Bacillus thuringiensis coated zinc oxide nanoparticle and its biopesticidal effects on the pulse beetle, *Callosobruchus maculatus*. J Photochem Photobiol B Biol. 2017;174:306-314. DOI: 10.1016/j.jphotobiol.2017.08.014

[49] Gouin S. Microencapsulation: industrial appraisal of existing technologies and trends. Trends Food Sci Technol. 2004;15:330-347. DOI: 10.1016/j.tifs.2003.10.005

[50] Otsuka H, Nagasaki Y, Kataoka K. PEGylated nanoparticles for biological and pharmaceutical applications. Adv Drug Deliv Rev. 2012;64:246-255. DOI: 10.1016/s0169-409x(02)00226-0

[51] Das S, Chakraborty J, Chatterjee S, Kumar H. Prospects of biosynthesized nanomaterials for the remediation of organic and inorganic environmental contaminants. Environ Sci: Nano. 2018;5:2784-2808. DOI: 10.1039/C8EN00799C

[52] Rai M, Ingle A. Role of nanotechnology in agriculture with special reference to management of insect pests. Appl Microbiol Biotechnol. 2012;94:287-293. DOI: 10.1007/s00253-012-3969-4

[53] Mochizuki H, Gautam PK, Sinha S, Kumar S. Increase Fertilizer and Pesticide Use Efficiency by Nanotechnology in Desert Afforestation, Arid Agriculture. Journal of Arid Land Studies. 2009;19(1):129-132

[54] Mali SC, Raj S, Trivedi R. Nanotechnology a novel approach to enhance crop productivity. Biochemistry and Biophysics Reports. 2020;24:100821. DOI: 10.1016/j.bbrep.2020.100821

[55] Rai V, Acharya S, Dey N. Implications of Nanobiosensors in Agriculture. Journal of Biomaterials and Nanotechnology. 2012;3(2):315-324. DOI: 10.4236/jbnb.2012.322039

[56] Garg D, Payasi DK. Nanomaterials in Agricultural Research: An Overview. In: Dasgupta N., Ranjan S., Lichtfouse E. (eds) Environmental Nanotechnology Volume 3. Environmental Chemistry for a Sustainable World, vol 27. Springer, Cham; 2020. DOI: 10.1007/978-3-030-26672-1_8

[57] Ghaffar N, Farrukh MA, Naz S. Applications of Nanobiosensors in Agriculture. In: Javad S. (eds) Nanoagronomy. Springer, Cham; 2020. DOI: 10.1007/978-3-030-41275-3_10

[58] Cicek S, Nadaroglu H. The use of nanotechnology in the agriculture. Advances in Nano Research. 2015;3(4), 207-223. DOI: 10.12989/anr.2015.3.4.207

[59] Hodson ME. The need for sustainable soil remediation. Elements. 2010;6(6):363-368. DOI: 10.2113/gselements.6.6.363

[60] Banwart S. Save our soils. Nature. 2011;474:151-152. DOI: 10.1038/474151a

[61] Abhilash PC, Dubey RK, Tripathi V, Srivastava P, Verma JP, Singh HB. Remediation and management of POPs-contaminated soils in a warming climate: Challenges and perspectives. Environ Sci Pollut Res. 2013;20(8):5879-5885. DOI: 10.1007/s11356-013-1808-5

[62] Gil-Díaz M, Ortiz LT, Costa G, Alonso J, Rodríguez-Membibre ML, Sánchez-Fortún S, Pérez-Sanz A, Martín M, Lobo MC. Immobilization and leaching of Pb and Zn in an acidic soil treated with zero-valent iron nanoparticles (nZVI): Physicochemical and toxicological analysis of leachates. Water Air Soil Pollut. 2014;225(6):1990. DOI: 10.1007/s11270-014-1990-1

[63] Enciu D, Toader A, Ursu I. Magnetic field nanosensor based on Mn impurities. Incas Bulletin. 2014;6(2):51-60. DOI: 10.13111/2066-8201.2014.6.2.5

[64] Bakshi M, Abhilash PC. Nanotechnology for soil remediation: Revitalizing the tarnished resource. In: Nano-Materials as Photocatalysts for Degradation of Environmental Pollutants. Elsevier. 2020;345-370. DOI: 10.1016/B978-0-12-818598-8.00017-1

[65] Yao Z, Li J, Xie H, Yu C. Review on remediation technologies of soil contaminated by heavy metals. Procedia Environ Sci. 2012;16:722-729. DOI: 10.1016/j.proenv.2012.10.099

[66] Sharma HD, Reddy KR. Geo-environmental engineering: site remediation, waste containment, and emerging waste management technologies. Wiley, Hoboken; 2004

[67] Rizwan MD, Singh M, Mitra CK, Morve RK. Eco-friendly application of nanomaterials: Nanobioremediation. J Nanopart. 2014;431787(7). DOI: 10.1155/2014/431787

[68] Raliya R, Saharan V, Dimkpa C, Biswas P. Nanofertilizer for precision and sustainable agriculture: current state and future perspectives. J Agric Food Chem. 2017;66(26):6487-6503. DOI: 10.1021/acs.jafc.7b02178

[69] Jiang D, Zeng G, Huang D, Chen M, Zhang C, Huang C. Remediation of contaminated soils by enhanced nano-scale zero valent iron. Environmental Research. 2010;163:217-227. DOI: 10.1056/NEJMoa1300955

[70] Cai C, Zhao M, Yu Z, Rong H, Zhang C. Utilization of nanomaterials for in-situ remediation of heavy metal (loid) contaminated sediments: A review. Sci Total Environ. 2019;662:205-217. DOI: 10.1056/NEJMoa1300955

[71] Trujillo -Reyes J, Peralta -Videa JR, Gardea -Torresdey JL. Supported and unsupported nanomaterials for water and soil remediation: are they a useful solution for worldwide pollution? J Hazard Mater. 2014;280:487-503. DOI: 10.1016/j.jhazmat.2014.08.029.

[72] Tang WW, Zeng GM, Gong JL, Liang J, Xu P, Zhang C. Impact of humic/fulvic acid on the removal of heavy metals from aqueous solutions using nanomaterials: A review. Sci Total Environ. 2014;468:1014-1027. DOI: 10.1016/j.scitotenv.2013.09.044

[73] Liang F, Fan J, Guo Y, Fan M, Wang J, Yang H. Reduction of nitrite by ultrasound-dispersed nano-scale zero-valent iron particles. Ind Eng Chem Res. 2008;47:8550-8554. DOI: 10.1021/ie8003946

[74] Elliott DW, Lien HL, Zhang WX. Degradation of lindane by zero-valent iron nanoparticles. J Environ Eng. 2009;135:317. DOI: 10.1061/(ASCE)0733-9372(2009)135:5(317)

[75] Shih YH, Tai YT. Reaction of decabrominated diphenyl ether by zero-valent iron nanoparticles. Chemosphere. 2010;78(10):1200-6. DOI: 10.1016/j.chemosphere.2009.12.061

[76] Araújoa R, Castrob ACM, Fiúza A. The use of nanoparticles in soil and water remediation processes. Mater Today: Proceedings. 2015;2:315-320. DOI: 10.1016/j.matpr.2015.04.055

[77] Raliya R, Tarafdar JC, Gulecha K, Choudhary K, Ram R, Mal P, Saran RP. Scope of nanoscience and nano technology in agriculture. J Appl Biol Biotechnol. 2013;1:41-44. DOI: 10.7324/JABB.2013.1307

[78] Hajirostamlo B, Mirsaeedghazi N, Arefnia M, Shariati M, Fard EA. The Role of Research and Development in Agriculture and Its Dependent Concepts in Agriculture. Asian J Appl Sci Eng. 2015;4:78-80.

[79] Pan B, Xing B. Applications and implications of manufactured nanoparticles in soils: A review. European Journal of Soil Science. 2012;63(4):437-456. DOI: 10.1111/j.1365-2389.2012.01475.x

[80] Wang P, Lombi E, Zhao FJ, Kopittke PM. Nanotechnology: A new opportunity in plant sciences. Trends in Plant Science. 2016;21(8):699-712. DOI: 10.1016/j.tplants.2016.04.005

[81] Fu R, Zhang X, Xu Z, Guo X, Bi D, Zhang W. Fast and highly efficient removal of chromium (VI) using humus-supported nano-scale zero-valent iron: Influencing factors, kinetics, and mechanism. Separation and Purification Technology. 2017;174:362-371. DOI: 10.1016/j.seppur.2016.10.058

[82] Elkady MF, Mahmoud MM, Rahman AEHM. Kinetic approach for cadmium sorption using microwave synthesized nano-hydroxyapatite. J Non-Cryst Solids. 2011;357(3):1118-1129. DOI: 10.1016/j.jnoncrysol.2010.10.021

[83] Cheng JM, Liu YZ, Wang HW. Effects of surface-modified nano-scale carbon black on Cu and Zn fractionations in contaminated soil. Int J Phytoremediation. 2014;16(1):86-94. DOI: 10.1080/15226514.2012.759530

[84] Luo HW, Wee LW, Wu YC. Hydrothermal synthesis of needle-like nanocrystalline zeolites from metakaolin and their applications for efficient removal of organic pollutants and heavy metals. Microporous and Mesoporous Materials. 2018;272:8-15. DOI: 10.1016/j.micromeso.2018.06.015

[85] Lin J, Su B, Sun M. Biosynthesized iron oxide nanoparticles used for optimized removal of cadmium with response surface methodology. Sci Total Environ. 2018;627:314-321. DOI: 10.1016/j.scitotenv.2018.01.170

[86] Fajardo C, Costa G, Nande M. Heavy metals immobilization capability of two iron-based nanoparticles (nZVI and Fe3O4): soil and freshwater bioassays to assess ecotoxicological impact. Sci Total Environ. 2019;656:421-432. DOI: 10.1016/j.scitotenv.2018.11.323

[87] Nel A, Xia T, Mädler L. Toxic potential of materials at the nanolevel. Science. 2006;311(5761):622-627. DOI: 10.1126/science.1114397

[88] Tang HQ, Xu M, Zhou XR. Acute toxicity and biodistribution of different sized copper nanoparticles in rats after oral administration. Mater Sci Eng C Mater Biol Appl. 2018;93:649-663. DOI: 10.1016/j.msec.2018.08.032

[89] Bigorgne E, Foucaud L, Lapied E. Ecotoxicological assessment of TiO2 byproducts on the earthworm *Eisenia fetida*. Environ Pollut. 2011;159(10): 2698-2705. DOI: 10.1016/j.envpol. 2011.05.024

[90] Li M, Yang Y, Xie JW. In-vivo and in-vitro tests to assess toxic mechanisms of nano ZnO to earthworms. Sci Total Environ. 2019;687:71-76. DOI: 10.1016/j. scitotenv.2019.05.476

[91] Nathanail CP, Gillett A, McCaffrey C, Nathanai J, Ogden R. A preliminary risk assessment protocol for renegade nanoparticles deployed during nanoremediation. Remed. J. 2016;26:95. DOI: 10.1002/rem.21471

[92] Cornelis G, Hund-Rinke K, Kuhlbusch T, Van den Brink N, Nickel C. Fate and bioavailability of engineered nanoparticles in soils: A review. Crit Rev Environ Sci Technol. 2014;44:2720-2764. DOI: 10.1080/10643389. 2013.829767

[93] Cornelis G, Ryan B, McLaughlin MJ, Kirby JK, Beak D, Chittleborough D. Solubility and batch retention of CeO2 nanoparticles in soils. Environ Sci Technol. 2011;45:2777-2782. DOI: 10.1021/es103769k

[94] Praetorius A, Labille J, Scheringer M, Thill A, Hungerbuhler K, Bottero JY. Heteroaggregation of titanium dioxide nanoparticles with model natural colloids under environmentally relevant conditions. Environ Sci Technol. 2014;48:10690-10698. DOI: 10.1021/es501655v

[95] Mukhopadhyay SS, Kaur N. Nanotechnology in Soil-Plant System. In: Kole C., Kumar D., Khodakovskaya M. (eds) Plant Nanotechnology. Springer, Cham; 2016. p. 329-348. DOI: 10.1007/978-3-319-42154-4_13

[96] DeRosa, MC, Monreal C, Schnitzer M, Walsh R, Sultan Y. Nanotechnology in fertilizers. Nat Nanotechnol. 2010;5(2):91. DOI: 10.1038/nnano.2010.2

[97] Tarafdar JC, Devakumar C, Chhonkar PK. National Academy of Agricultural Sciences Policy Paper#63 Nanotechnology in Agriculture: scope and current relevance. National Academy of Agricultural Sciences New Delhi India; 2013. p. 20.

[98] Leggo PJ. An investigation of plant growth in an organo-zeolitic substrate and its ecological significance. Plant Soil. 2000:219(1):135-146. DOI: 10.1023/A:1004744612234.

[99] Millán, Guillermo, Agosto, Florencia, Vázquez, Mabel, Botto, Lia, Lombardi, Luciano, Juan, Luciano. Uso de clinoptilolita como un vehículo de fertilizantes nitrogenados en un suelo de la región Pampeana de Argentina. Ciencia e investigación agrarian. 2008;35(3):293-302. DOI: 10.4067/ S0718-16202008000300007

[100] Chinnamuthu CR, Boopathi PM. Nanotechnology and agroecosystem. Madras Agric J. 2009;96:17-31

[101] Jinghua G. Synchrotron radiation, soft X-ray spectroscopy and nano-materials. Int J Nanotechnol. 2004;1:193-225. DOI: 10.1504/ IJNT.2004.003729

[102] You T, Liu D, Chen J, Yang Z, Dou R, Gao X, Wang L. Effects of metal oxide nanoparticles on soil enzyme activities and bacterial communities in two different soil types. J Soils Sediments. 2018;18:211-221. DOI: 10.1007/s11368-017-1716-2

[103] Cao J, Feng Y, Lin X, Wang J. Arbuscular mycorrhizal fungi alleviate the negative effects of iron oxide nanoparticles on bacterial community in rhizospheric soils. Front Environ Sci. 2016;4(10). DOI: 10.3389/fenvs.2016. 00010

[104] Jo YK, Kim BH, Jung G. Antifungal activity of silver ions and nanoparticles on phytopathogenic fungi. Plant Dis. 2009;93:1037-1043. DOI: 10.1094/PDIS-93-10-1037

[105] Bryaskova R, Pencheva D, Nikolov S, Kantardjiev T. Synthesis and comparative study on the antimicrobial activity of hybrid materials based on silver nanoparticles (AgNps) stabilized by polyvinylpyrrolidone (PVP). J Chem Biol. 2011;4:185-191. DOI: 10.1007/s12154-011-0063-9

[106] Tayel AA, EL-TRAS WF, Moussa S, EL-BAZ AF, Mahrous H, Salem MF, Brimer L. Antibacterial action of zinc oxide nanoparticles against foodborne pathogens. J Food Saf. 2011;31:211-218. DOI: 10.1111/j.1745-4565.2010.00287.x

[107] Siddiqui MH, Al-Whaibi MH. Role of nano-SiO2 in germination of tomato (Lycopersicum esculentum seeds Mill.). Saudi Journal of Biological Sciences. 2014;21(1):13-17. DOI: 10.1016/j.sjbs.2013.04.005

[108] Siddiqui MH, Al-Whaibi MH, Faisal M, Al Sahli AA. Nano-silicon dioxide mitigates the adverse effects of salt stress on *Cucurbita pepo* L. Environmental Toxicology and Chemistry. 2014;33(11):2429-37. DOI: 10.1002/etc.2697

[109] Hasanuzzaman M, Fujita M. Selenium pretreatment upregulates the antioxidant defense and methylglyoxal detoxification system and confers enhanced tolerance to drought stress in rapeseed seedlings. Bio Trace Elem Res. 2011;143(3):1758-76. DOI: 10.1007/s12011-011-8998-9

[110] Zohra E, Ikram M, Omar AA, Hussain M, Satti SH, Raja NI, Mashwani Z, Ehsan M. Potential applications of biogenic selenium nanoparticles in alleviating biotic and abiotic stresses in plants: A comprehensive insight on the mechanistic approach and future perspectives. Green Processing and Synthesis. 2021;10(1):456-75. DOI: 10.1515/gps-2021-0047

[111] Ashkavand P, Tabari M, Zarafshar M, Tomášková I, Struve D. Effect of SiO2 Nanoparticles on Drought Resistance in Hawthorn Seedlings. Forest Research Papers. 2015;76:350-359.

[112] Bose P. How can Nanotechnology Help Mitigate Environmental Stress on Crop Plants? [Internet]. 2020. Available from: https://www.azonano.com/article.aspx?ArticleID=5502 [Accessed: 2021-10-08]

[113] Abdel-Aziz H.M.M, Hasaneen MNA, Omer AM. Nano-chitosan-NPK fertilizer enhances the growth and productivity of wheat plants grown in sandy soil. Spanish. J. Agric. Res. 2016;14:1-9. DOI: 10.5424/sjar/2016141-8205

[114] Chahardoli A, Karimi N, Ma X, Qalekhani F. Effects of engineered aluminum and nickel oxide nanoparticles on the growth and antioxidant defense systems of Nigella arvensis L. Scientific Reports. 2020;10(1):3847. DOI: 10.1038/s41598-020-60841-6

[115] Rajeshwari A, Kavitha S, Alex SA, Kumar D, Mukherjee A, Chandrasekaran N, Mukherjee A. Cytotoxicity of aluminum oxide nanoparticles on *Allium cepa* root tip--effects of oxidative stress generation and biouptake. Environmental Science and Pollution Research. 2015;22(14):11057-66. DOI: 10.1007/s11356-015-4355-4.

[116] Kaveh R, Li YS, Ranjbar S, Tehrani R, Brueck CL, Van Aken B. Changes in *Arabidopsis thaliana* gene expression in response to silver nanoparticles and silver ions. Environmental Science & Technology.

2013;47(18):10637-44. DOI: 10.1021/es402209w

[117] Rico CM, Peralta-Videa JR, Gardea-Torresdey JL. Chemistry, biochemistry of nanoparticles, and their role in antioxidant defense system in plants. In: Siddiqui M, Al-Whaibi M, Mohammad F. (eds) Nanotechnology and plant sciences. Springer, Cham; 2015. p. 1-17. DOI: 10.1007/978-3-319-14502-0_1

[118] Chandra S, Chakraborty N, Dasgupta A, Sarkar J, Panda K, Acharya K. Chitosan nanoparticles: a positive modulator of innate immune responses in plants. Sci Rep. 2015;5: 15195. DOI: 10.1038/srep15195

[119] Hong F, Yang F, Liu C, Gao Q, Wan Z, Gu F, Wu C, Ma Z, Zhou J, Yang P. Influences of nano-TiO2 on the chloroplast aging of spinach under light. Biol Trace Elem Res. 2005;104(3):249-260. DOI: 10.1385/BTER:104:3:249

[120] Tan XM, Lin C, Fugetsu B. Studies on toxicity of multi-walled carbon nanotubes on suspension rice cells. Carbon. 2009;47:3479-3487. DOI: 10.1016/j.carbon.2009.08.018

[121] Begum P, Fugetsu B. Phytotoxicity of multi-walled carbon nanotubes on red spinach (*Amaranthus tricolor* L.) and the role of ascorbic acid as an antioxidant. J Hazard Mater. 2012;243: 212-222. DOI: 10.1016/j.jhazmat.2012.10.025

[122] Rangaraj SR, Gopalu K, Muthusamy P, Rathinam Y, Venkatachalam R, Narayanasamy K. Augmented biocontrol action of silica nanoparticles and *Pseudomonas fluorescens* bioformulant in maize (Zea mays L.). RSC Adv. 2014;4:8461-8465. DOI: 10.1039/C3RA46251J

[123] Nekrasova GF, Ushakova OS, Ermakov AE, Uimin MA, Byzov IV. Effects of copper (II) ions and copper oxide nanoparticles on *Elodea densa* Planch. Russ J Ecol. 2011;42:458-463. DOI: 10.1134/S1067413611060117

[124] Majumdar S, Peralta-Videa JR, Bandyopadhyay S, Castillo-Michel H, Hernandez-Viezcas JA, Sahi S, Gardea-Torresdey JL. Exposure of cerium oxide nanoparticles to kidney bean shows disturbance in the plant defense mechanisms. J Hazard Mater. 2014;278:279-287. DOI: 10.1016/j.jhazmat.2014.06.009

[125] Zhao LJ, Hernandez-Viezcas JA, Peralta-Videa JR, Bandyopadhyay S, Peng B, Munoz B, Keller AA, Gardea-Torresdey JL. ZnO nanoparticle fate in soil and zinc bioaccumulation in corn plants (Zea mays) influenced by alginate. Environ Sci Process Impacts. 2013;15:260-266. DOI: 10.1039/c2em30610g

[126] Arif N, Yadav V, Singh S, Kushwaha BK, Singh S, Tripathi DK, Vishwakarma K, Sharma S, Dubey NK, Chauhan DK. Assessment of antioxidant potential of plants in response to heavy metals. In: Singh A, Prasad S, Singh R. (eds) Plant responses to xenobiotics. Springer Singapore, Singapore; 2016. p. 97-125. DOI: 10.1007/978-981-10-2860-1_5

[127] Anonymous. FAO's Director-General on How to Feed the World in 2050. Population and Development Review. 2009;35(4):837-839. DOI: 10.1111/j.1728-4457.2009.00312.x

[128] Weir A, Westerhoff P, Fabricius L, Hristovski K, von Goetz N. Titanium Dioxide Nanoparticles in Food and Personal Care Products. Environmental Science & Technology. 2012;46(4):2242-2250. DOI: 10.1021/es204168d

[129] Cushen M, Kerry J, Morris M, Cruz-Romero M, Cummins E. Nanotechnologies in the food industry – Recent developments, risks and regulation. Trends in Food Science &

Technology. 2012;24(1):30-46. DOI: 10.1016/j.tifs.2011.10.006

[130] Ezhilarasi P, Karthik P, Chhanwal N, Anandharamakrishnan C. Nanoencapsulation Techniques for Food Bioactive Components: A Review. Food and Bioprocess Technology. 2012;6(3): 628-647. DOI: 10.1007/s11947-012-0944-0

[131] Ranjan S, Dasgupta N, Chakraborty AR, Samuel SM, Ramalingam C, Shanker R, Kumar A. Nanoscience and nanotechnologies in food industries: opportunities and research trends. Journal of Nanoparticle Research. 2014;16(6):2464. DOI: 10.1007/s11051-014-2464-5

[132] An J, Zhang M, Wang S, Tang J. Physical, chemical and microbiological changes in stored green asparagus spears as affected by coating of silver nanoparticles-PVP. LWT - Food Science and Technology. 2008;41(6):1100-1107. DOI: 10.1016/j.lwt.2007.06.019

[133] Mao X, Huang J, Leung M, Du Z, Ma L, Huang Z. Novel Core-Shell Nanoparticles and Their Application in High-Capacity Immobilization of Enzymes. Applied Biochemistry and Biotechnology. 2006;135(3):229-246. DOI: 10.1385/abab:135:3:229

[134] Huang Q, Yu H, Ru Q. Bioavailability and Delivery of Nutraceuticals Using Nanotechnology. Journal of Food Science. 2010;75(1):R50-R57. DOI: 10.1111/j.1750-3841.2009.01457.x

[135] Ozimek L, Pospiech E, Narine S. Nanotechnologies in food and meat processing. Acta Sci Pol Technol Aliment. 2010;9(4):401-412

[136] Rashidi L, Khosravi-Darani K. The Applications of Nanotechnology in Food Industry. Crit Rev Food Sci Nutr. 2011;51(8):723-30. DOI: 10.1080/10408391003785417

[137] Armstrong G, Armstrong A, Canales J, Bruce P. Nanotubes with the TiO2-B structure. Chemical Communications. 2005;(19):2454. DOI: 10.1039/b501883h

[138] Rao M, Jha A, Prasad K. Nanomaterials: An Upcoming Fortune to Waste Recycling. In: Prasad R, Jha A, Prasad K. (eds) Exploring the Realms of Nature for Nanosynthesis. Nanotechnology in the Life Sciences. Springer, Cham; 2018. p. 241-271. DOI: 10.1007/978-3-319-99570-0_11

[139] Baruah S, Dutta J. Nanotechnology applications in pollution sensing and degradation in agriculture: a review. Environ Chem Lett. 2009;7(3):191-204. DOI: 10.1007/s10311-009-0228-8

[140] Alvarado MA, Guzmán ON, Solís NM, Vega-Baudrit J. Recycling and Elimination of Wastes obtained from Agriculture by using Nanotechnology: Nanosensors. Int J Biosen Bioelectron. 2017;3(5):368-375. DOI: 10.15406/ijbsbe.2017.03.00084

[141] Dutta N, Mukhopadhyay A, Dasgupta A, Chakrabarti K. Improved production of reducing sugars from rice husk and rice straw using bacterial cellulase and xylanase activated with hydroxyapatite nanoparticles. Bioresource Technology. 2014;153:269-277. DOI: 10.1016/j.biortech.2013.12.016

[142] Nogueira V, Lopes I, Freitas A, Rocha-Santos T, Gonçalves F, Duarte A, Pereira R. Biological treatment with fungi of olive mill wastewater pre-treated by photocatalytic oxidation with nanomaterials. Ecotoxicology and Environmental Safety. 2015;115:234-242. DOI: 10.1016/j.ecoenv.2015.02.028

[143] Karimi M, Keyhani A, Akram A, Rahman M, Jenkins B, Stroeve P. Hybrid response surface methodology-genetic algorithm optimization of ultrasound-assisted transesterification of waste oil catalysed by immobilized lipase on

mesoporous silica/iron oxide magnetic core-shell nanoparticles. Environmental Technology. 2013;34(13-14):2201-2211. DOI: 10.1080/09593330.2013.837939

[144] Yu C, Huang L, Kuan I, Lee S. Optimized Production of Biodiesel from Waste Cooking Oil by Lipase Immobilized on Magnetic Nanoparticles. International Journal of Molecular Sciences. 2013;14(12):24074-24086. DOI: 10.3390/ijms141224074

[145] Alhassan F, Rashid U, Taufiq-Yap Y. Synthesis of Waste Cooking Oil Based Biodiesel via Ferric-Manganese Promoted Molybdenum Oxide / Zirconia Nanoparticle Solid acid Catalyst: Influence of Ferric and Manganese Dopants. Journal of Oleo Science. 2015;64(5):505-514. DOI: 10.5650/jos.ess14228

[146] Arshadi M, Foroughifard S, Etemad Gholtash J, Abbaspourrad A. Preparation of iron nanoparticles-loaded *Spondias purpurea* seed waste as an excellent adsorbent for removal of phosphate from synthetic and natural waters. Journal of Colloid and Interface Science. 2015;452:69-77. DOI: 10.1016/j.jcis.2015.04.019

[147] Ahmad N, Sharma S. Green Synthesis of Silver Nanoparticles Using Extracts of *Ananas comosus*. Green and Sustainable Chemistry. 2012;2(4):141-147. DOI: 10.4236/gsc.2012.24020

[148] Sinsinwar S, Sarkar M, Suriya K, Nithyanand P, Vadivel V. Use of agricultural waste (coconut shell) for the synthesis of silver nanoparticles and evaluation of their antibacterial activity against selected human pathogens. Microbial Pathogenesis. 2018;124:30-37. DOI: 10.1016/j.micpath.2018.08.025

[149] Bose D, Chatterjee S. Biogenic synthesis of silver nanoparticles using guava (*Psidium guajava*) leaf extract and its antibacterial activity against *Pseudomonas aeruginosa*. Applied Nanoscience. 2015;6(6):895-901. DOI: 10.1007/s13204-015-0496-5

[150] Mishra A, Sardar M. Rapid Biosynthesis of Silver Nanoparticles Using Sugarcane Bagasse — An Industrial Waste. Journal of Nanoengineering and Nanomanufacturing. 2013;3(3):217-219. DOI: 10.1166/jnan.2013.1135

[151] Vijayakumar S, Vaseeharan B, Malaikozhundan B, Gopi N, Ekambaram P, Pachaiappan R. Therapeutic effects of gold nanoparticles synthesized using Musa paradisiaca peel extract against multiple antibiotic resistant *Enterococcus faecalis* biofilms and human lung cancer cells (A549). Microbial Pathogenesis. 2017;102:173-183. DOI: 10.1016/j.micpath.2016.11.029

[152] Krishnaswamy K, Vali H, Orsat V. Value-adding to grape waste: Green synthesis of gold nanoparticles. Journal of Food Engineering. 2014;142:210-220. DOI: 10.1016/j.jfoodeng.2014.06.014

[153] Molins R. Oportunidades y amenazas de la nanotecnología para la salud, los alimentos, la agricultura y el ambiente. Perspectivas. 2008;4:38-52

[154] Mousavi SR, Rezaei M. Nanotechnology in agriculture and food production. J Appl Environ Biol Sci. 2011;1:414-419

[155] Mishra VK, Dwivedi DK, Mishra UK. Emerging consequence of nanotechnology in agriculture: an outline. Trends Biosci. 2013;6(5):503-506

[156] Baruwati B, Varma RS. High value products from waste: grape pomace extract-a three-in-one package for the synthesis of metal nanoparticles. ChemSusChem. 2009;2(11):1041-1044. DOI: 10.1002/cssc.200900220

[157] Zhang XL, Tyagi YRDS, Surampalli RY. Biodiesel production

from heterotrophic microalgae through transesterification and nanotechnology application in the production. Renew Sustain Energy Rev. 2013;26:216-223. DOI: 10.1016/j.rser.2013.05.061

[158] Obi FO, Ugwuishiwu BU, Nwakaire JN. Agricultural waste concept, generation, utilization and management. Nigerian Journal of Technology. 2016;35:957-964. DOI: 10.4314/njt.v35i4.34

[159] Romanovski V. Agricultural waste based-nanomaterials: green technology for water purification. Aquananotechnology. 2021;577-595. DOI: 10.1016/b978-0-12-821141-0.00013-6

[160] FAO. Food wastage footprint: Impacts on natural resources. FAO, Rome; 2013.

[161] Javad S, Akhtar I, Naz S. Nanomaterials and Agrowaste. Nanoagronomy. 2020;197-207. DOI: 10.1007/978-3-030-41275-3_11

Chapter 2

Enhancing the Productivity of Field Crops through Nano-Fertilizer

Rahul Sadhukhan, L. Devarishi Sharma, Suman Sen, Snehashis Karmakar, Koushik Banerjee and Kirtiranjan Baral

Abstract

The growth of agricultural sectors can be maintained by increasing crop productivity through soil, water, and nutrient management. The most important management practice is nutrient management, which is supported by the effective use of nano-technology, especially nano-fertilizers. It is well known that nano-fertilizers are nutrient carriers of nano dimensions ranging from 30 to 40 nm (10^{-9} m or one-billionth of a meter). Due to their high surface area, they can hold abundant nutrients ions and release them slowly and steadily, commensurate with crop demand. Nano-fertilizers are easily uptaken and assimilated by the plants because of their ease of solubility, stability, controlled release in time, and easy mode of delivery and disposal. Due to nano fertilizers characteristics, different commercial products are available in the market, namely *Nanogro*, *Geohumus*, *NanoGreen*, and *Lithovit High Yield fertilizer*, which can be demonstrated among the farmers for increasing agricultural performance through soil and nutrient management. Besides, nano-fertilizer has good criteria like disease resistance properties. Nanoparticles of ZnO, CuO, and MgO can kill different fungal infections of crop plants. Though nano-fertilizers can be beneficial for improving agricultural performance, it has a detrimental effect on soil microflora, fauna, animals, and humans. It is associated with several diseases or hazards like high blood pressure, blood clots, stroke, arrhythmia, heart disease, etc. Nano-fertilizer also improves the yield of several field crops like pearl millet, wheat, pomegranate, onion, tomato, soybean, and vegetable crops like spinach and cucumber. Nano fertilizers also have sound capabilities to find the solution against the issues arising in modern agriculture due to conventional fertilizer application. Thus, nano-fertilizer has the potential to improve the yield of several field crops.

Keywords: nano-fertilizers, commercial products, fungal infection, disease, disease resistance, surface area, cereal and vegetables, solubility, stability and controlled release

1. Introduction

Increasing population growth and shortage of available land & water resources are significant concerns for food-saving agriculture. Potential agricultural growth can be achieved through productivity improvement through soil, water, and nutrients management, assisted by the successful use of new technologies such as nano-fertilizers. Nano-fertilizers are those preparations with nano-dimensional nutrient carriers ranging from 30 and 40 nm (10^{-9} m or one billionth of a meter). They can retain sufficient nutrient ions because of their large specific surface area and slowly or gradually release them in exact amounts meeting crop requirements [1]. Compared with bulk fertilizers, it has a high specific surface area, small size, and reactivity of nano-fertilizers can increase the diffusion, solubility, and nutrient availability to plants and boost agricultural productivity. Using fertilizer carriers for the construction of smart fertilizers, nano-fertilizer have presented the feasibility of exploring nano-structured materials as new facilities to increase the performance of nutrient usage and minimize environmental degradation. Micronutrients such as iron, manganese, zinc, and copper are becoming factors that improve yield and are primarily responsible for making the standard nutritional quality of food items. Once applied to the soils, they respond quickly, produce chemical precipitates, soils' organo-mineral matrix, and react with clay colloids. Micro-nutrients are substantially lost due to leaching in high rainfall regions. Thereby, the efficiency of micronutrient usage (MUE) is <5%. Micronutrient-based nano-fertilizers can increase the accessibility of particular micronutrients to crops and improve agricultural productivity [2]. Since the production and implementation of nano-fertilizers are still at an early stage, certain specific reports are published on the effects and benefits of applying micronutrient-based nano-fertilizers in the field. Any discovery in new innovative technologies to enhance agricultural productivity and the supply of nutrients might be a landmark in nano-technology research and at the beginning of the next Green Revolution. So, the main objectives of this study are

i. To study the significance of nano-fertilizer in crop nutrition

ii. To understand the properties and methodologies of nano-fertilizer preparation

iii. To classify the different kinds of nano-fertilizers

iv. To understand about the commercial nano-formulation

v. To increase the productivity of agricultural crops

vi. To investigate the harmful effect of nano-particle related fertilizers

Indiscriminate and massive application of fertilizers leads to the formation of a non-porous layer between the soil particles, which has a negative effect on the agricultural soil and leads to a rise in the groundwater level and salinity as well and ultimately it causes the death of the roots of the field crops by reducing the absorption of nutrient by the plants that are essential in the crop nutrition. But nano-fertilizers cut the barriers of that type of problem in agriculture [3].

Recently, our modern agriculture has been habituated by the use of high rates of chemical fertilizer. For example, the global production of chemical fertilizer is about 188.2 Mt. in 2019. It is also expected that chemical fertilizer consumption can also increase by double to feed the 9.6 billion population by 2050. Besides, applications of conventional fertilizers have low nutrient use efficiency, a high risk of environmental pollution, and a possible risk of breaking the food chain by destroying the ecosystem [4]. So, sustainable agriculture is a new approach for the solution of recent problems. So, implementing new innovative techniques like nano-fertilizers into the farmers field may solve fertilizer application problems. That's why nano-fertilizers are gaining importance day to day to agriculture. It has several advantages, like it releases nutrients according to plant requirements. It may replace several pesticides as it has disease-resistant properties (ZnO, MgO, and CuO-based nano fertilizers).

We know that chemical fertilizers release the nutrients in 4–10 days, whereas nano-fertilizer releases their nutrients in 40–50 days; as a result, nitrogenous fertilizer from the conventional system are lost rapidly from the field through volatilization, leaching, and run-off. But we can minimize this problem by applying nano-fertilizer and improving N's nutrient use efficiency in the field [4].

2. Significance of nano-fertilizers in crop nutrition

1. Strong Stability and Solubility

2. Solid effectiveness

3. Controlled release of the nano fertilizer in time

4. Improved targeted behavior with the right concentration

5. Reduced eco-toxicity

6. Secure, uncomplicated delivery and disposal mode

2.1 Probable mode of cellular uptake of nanoparticles in a plant cell

Nano-fertilizers get into the plant system through a mechanism [5]. The mechanism is expressed in the **Figure 1**.

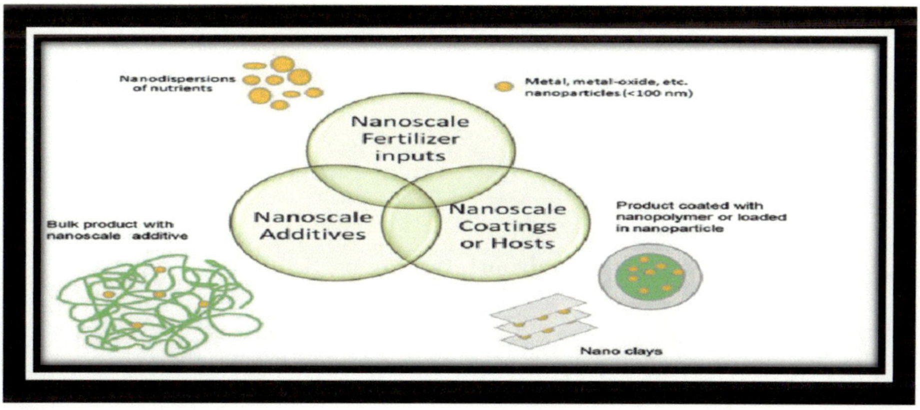

Figure 1.
Composition of nanoparticles.

3. Nano-fertilizers delivery systems

Nanoparticles can provide nutrients to particular target sites in living organisms. The preparation of nutrients based nano-particles is typically carried out by:

1. Nano-particles absorption.
2. Ligands mediated nano-particles attachment.
3. The formation of the nano-particulate polymeric shell through encapsulation.
4. Polymeric entrapment of nano-particles.
5. Self-possessed synthesis of nano-particles.

4. Principles of nano-fertilizers preparation

The dissolution kinetics of nano-particle by The Scientist Noyes-Whitney.

The dissolution rate of nano-particles of the nano-fertilizers is directly proportional to their surface area because a relatively larger interface for dissolution is available, which promotes the dissolution of dissolved ions away from the particle.

According to *the Ostwald-Freundlich principle.*

The relative solubility of a spherical particle of the same material increases as the particle size decreases in the solid-liquid system—the solubility of the particle increases when the equivalent spherical diameter is <0.1 um.

5. Properties of polymer encapsulating nano-particle

1. Rapid-release: on contact with a surface, the capsule shell breaks (e.g., when fertilizer hits a leaf) [6]
2. Unique release: when a molecular receptor binds to a specific chemical, the shell is built to break open
3. Moisture release: in the presence of water, the shell breaks down and releases material (e.g., in soil)
4. Thermal release: the shell only releases components as the atmosphere heats over a certain temperature.
5. pH release: nanocapsule only disintegrates in particular acid or alkaline environments.
6. Slow-release: over a longer period, the capsule slowly releases its payload (e.g., for slow delivery of a substance in the fields such as nano-coated urea).

5.1 Unique properties of nano-particles (NPs)

a. Very high specific surface area (**Figure 2** and **Table 1**)

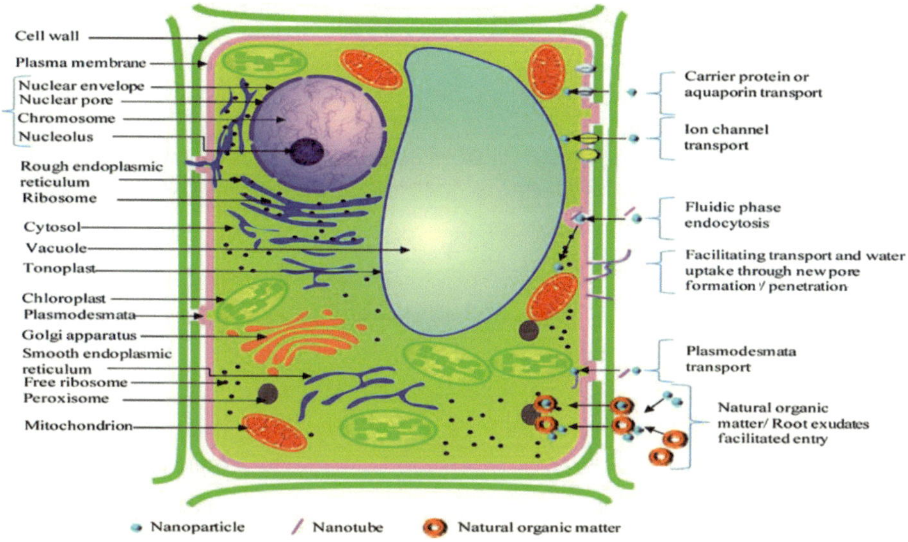

Figure 2.
Probable mode of cellular uptake of nanoparticles in a plant cell [5].

Properties	Nano-fertilizer enabled technology
Solubility and nano-sized formulation of mineral micronutrients (MN)	Nano-sized formulation of MN improve solubility and dispersion of insoluble nutrients in soil, reduce soil absorption and fixation and increase the bioavailability
Nutrient uptake efficiency	It increase fertilizer efficiency and uptake ratio of the soil nutrients in crop production and save fertilizer
Controlled release modes	Controlled release through encapsulation in envelope forms of semipermeable membrane coated by resin-polymer waxes
Effective duration of nutrient release	It extends effective duration of nutrient supply of fertilizers in to soil
Loss rate of fertilizer nutrients	Reduce nutrient loss by reducing leaching

Table 1.
Properties of various kinds of nano-fertilizers.

b. High surface energy

c. Quantum confinement

5.2 Classification of NPs

(a) Carbonaceous NPs, (b) metal oxide NPs, (c) zero valent metals and (d) nano-polymers (**Figure 3** and e.g., **Tables 2–4**).

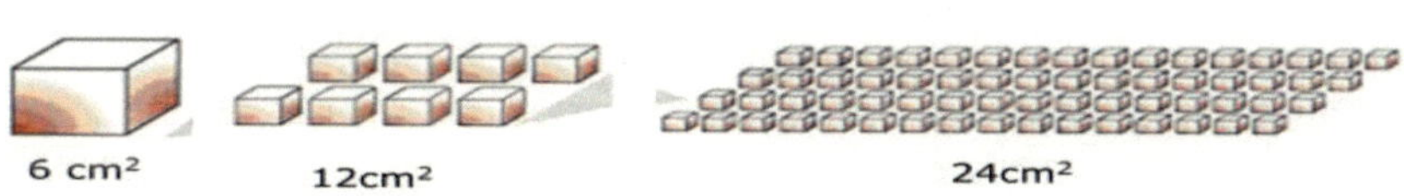

Figure 3.
Novel properties of nanoparticles: surface area per unit weight increases, so more atoms come on the surface to interact occurred in nanoparticles.

Nano contents	Applications
Ammonium salts (prepared through physical methods)	Ammonium magnesium phosphate containing nano-composites improved stability and slow release properties
Nano scale fertilizer particle (urea, KCl and NH_4Cl are prepared through emulsification, coating and shearing process)	Those help in slow and controlled release of nutrients
ZnO as agricultural trace element fertilizers	ZnO powder mixed with polymeric wetting agents and cellulose based thickening agent to efficiently uptake by plants

Table 2.
Nano-scale fertilizer inputs.

Nano contents	Applications
Bentonite clay	Nanoscale compound fertilizer additives bind soils and helps in N-fixing bacteria for improving soil fertility.
Humic acid and talc powder	Si nano-particle mixed with humic acid increases water and mineral holding capacity of soil. Reduced fertilizer use and improved crop yield
Rare earth materials (REMs)	REMs nanoparticles mixed with MgO, N, P and K improved soil fertility and control release of nutrients

Table 3.
Nano-scale fertilizer additives.

Nano contents	Applications
Halloysite	Halloysite and other nanotubes (Dolomites) mixed nano-materials used as controlled release fertilizers.
Phosphate fertilizer containing nano-hydroxy apatite	Those help in maintaining P level in soil and reduce its loss through water.
Attapulgite	N, P and K fertilizers embedded in the nano-pores of attapulgite clay yield slow release of fertilizer nutrient.

Table 4.
Nano-scale films and host materials.

6. Preparation of coated nano fertilizer granules

i. MAP & commercial urea granules sieved to obtain 2000–3350 um and 1676–2000 um size, respectively.

ii. Coating of MAP and Urea granules by Zn @ 1.5% by weight (by adding powder of ZnO nano particles.

iii. Spraying of ultrapure deionized water and use of nebulizer to provide good blending agent (50 uL/min) to ZnO powder.

iv. Coated granules are air-dried at 30% RH in the laminar flow.

7. Biosynthesis of Zn nano-fertilizers in laboratory

i. The Fungi, *Rhizoctonia bataticola* TFR 6 (strain) taken in 250 ml Erlenmeyer flak with 100 mL Potato dextrose Broth medium (pH adjusted to 8.5) MAP= Mono Ammonium Phosphate, NUE= Nitrogen Use Efficiency, TFR 6 is a strain

ii. After incubation, fungal mycelia separated from culture by filtration with Whatman No.1 paper

iii. Harvested mycelia kept in a rotary shaker at 150 rpm with 28°C for 12 h

iv. Cell-free filtrates + salt solution of ZnO with 0.1 mM concentration help synthesize Zn nanoparticles.

8. Review of literature

^{15}N studies using maize as a model system revealed that NUE from nano-fertilizer was higher than conventional fertilizer urea [7] (**Figure 4**).

Subramanian and Sharmila Rahale [7] reported that 15 N studies which were used in maize as a model system revealed that the N use efficiency from nano-fertilizer was 82%, and the conventional fertilizer (urea) registered 42% with a net higher nitrogen use efficiency of 40%, which is hardly achievable in the conventional systems.

Bakhtiari et al. [8] also experimented during 2015 to assess the effect of Fe nano particles on yields and quality of wheat. They found that foliar spray of Fe @ 0.04%

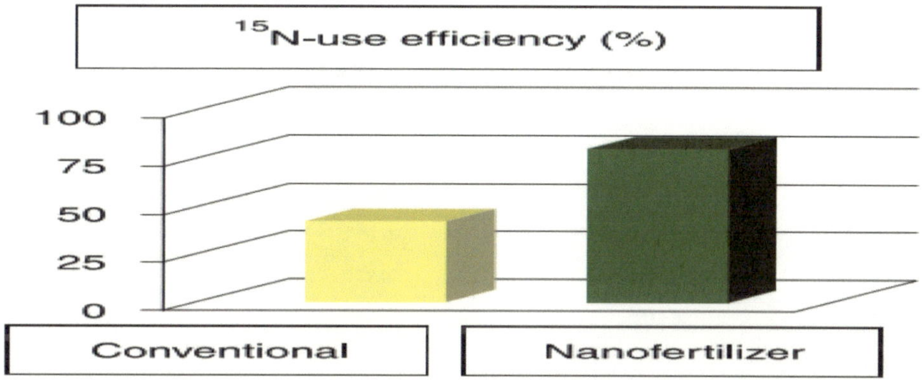

Figure 4.
Comparison of N based nano-fertilizers with conventional N fertilizers.

Concentrations (%) of Fe nano fertilizers	Spike Weight (g)	1000 grain wt (g)	Biological yields (kg/ha)	Grain yields (kg/ha)	Protein content (%)
0	536.33[c]	32.82[d]	8320.0[bc]	3316.5[c]	13.77[c]
0.01	561.33[c]	34.12[dc]	8520.0[ab]	3421.5[bc]	15.51[b]
0.02	604.21[b]	35.46[bc]	8620.0[ab]	3506.5[abc]	15.86[b]
0.03	647.96[a]	36.49[ab]	8845.0[a]	3689.0[ab]	16.59[a]
0.04	**666.96[a]**	**37.96[a]**	**8895.0[a]**	**3776.5[a]**	**16.44[a]**

[a,b,c] *is used in DMRT test for statistics either at par or different*

Table 5.
Effect of Fe based nano-fertilizers on wheat crop.

enhanced the spike weight, total biological yields, grain yields, and protein content in grains (**Table 5**).

Dey et al. [2] explained the various effects of nano-particles on plant growth (**Table 6**).

Davarpanah et al. [9] experimented during the year 2014–2015 to study the effect of nano-Zn and nano-B foliar fertilization on fruit diameter and fruit yield of pomegranate. They concluded that foliar spray of Boron and Zinc @ 4.5 and 120 ppm resulted in higher fruit diameter and fruit yield compared to other treatments. Where B1, B2 and Zn1, Zn2 are 3.25, 6.5 mg/L B and 60, 120 mg/L Zn, respectively (**Table 7**).

Likely, Tarafdar et al. [10] conducted a field experiment about nano-fertilizer on pearl millet to study the effect of Zn nano-fertilizers on pearl millet var. HHB 67. It was found that nano-fertilizer (Zn source) responded as well as a field application

Nanoparticle	Effect on plant growth
Carbon nanotubes (CNTs)	Improve root growth of onion and cucumber
Nano-Si	Develops salinity stress on tomato seed germination
1000 ppm nano-ZnO (25 nm)	Results highest chlorophyll content, higher seedling vigor, early vegetative growth and significant pod yield of peanut
Nano-SiO_2 and TiO_2	Hastens germination and growth in soybean
Nanoscale titanium dioxide (TiO_2)	Encourages photosynthesis and growth of spinach

Table 6.
Nano particles for better seed germination and plant growth [2].

Treatments	No. of fruits/tree	Pomegranate Fruit dia (mm)	Yields (kg/tree)
Control	50.6[d]	75.5[abc]	13.8[e]
Zn1 + B0	52.7[cd]	76.5[abc]	14.3[de]
Zn1 + B1	51.3[d]	78.2[ab]	15.0[cd]
Zn1 + B2	65.9[a]	76.6[abc]	18.5[a]
Zn2 + B1	58.7[b]	74.9[abc]	16.2[b]
Zn2 + B2	**63.0[ab]**	**78.8[a]**	**18.4[a]**
Significance	S	S	S

Table 7.
Effect of nano-Zn and nano-boron fertilizer combination on pomegranate crop.

Treatments	Pearl millet Grain yield ($kg\ ha^{-1}$)	Dry biomass ($kg\ ha^{-1}$)	Zn concentration ($mg\ kg^{-1}$)
Control	1065	5192	35.5
Ordinary Zn	1217	5214	39.8
Nano Zn	1467	5841	39.2
CV	48	142	3.1
LSD (p = 0.05)	17.6	52.2	1.1

Table 8.
Performance of nano-Zn fertilizer on pearl millet crop.

on pearl millet. The pearl millet yield and biomass yield increased by nano-fertilizer over ordinary fertilizers (**Table 8**).

Metal oxide nanomaterials, such as CuO, ZnO, and MgO, could also effectively control many plants and soil-borne diseases caused by *Botrytis cinerea*, *Alternaria alternate*, *Monilinia fructicola*, *Colletotrichum gloeosporioides*, *Fusarium solani*, *Fusarium oxysporum* fsp Radicis Lycopersici, *Verticillium dahliae*, *Phytophthora infestans* and *Ralstonia solanacearum* in many plant species [11].

8.1 Nano-fertilizers in precision agriculture

Nano fertilizer has significant advantages over conventional fertilizers (**Table 9**).

8.2 Delivery of fertilizer

Nano-particles hold nutrients more intensely due to high surface tension and provide high surface protection of surface particles.

8.2.1 Chemical fertilizers

Chemical N-fertilizers cause atmospheric N_2O emission: Solution—nano-coated urea, CRF (Zeolite, Halloysite, and Montmorillonite use for controlled release of nitrogenous fertilizers).

Properties	Nano fertilizers	Conventional fertilizers
Solubility and dispersion of nutrients	It improves solubility and dispersion of insoluble mineral nutrients in soil and make bioavailability to plants	Less bioavailable to plants due to less solubility and larger particle size
Nutrient uptake efficiency	It increases fertilizer use efficiency and uptake ratio of soil nutrient by plants and saves fertilizers	Due to its bulk composites not efficiently uptake by plants and reduces efficiency
Control release modes	Nutrients release precisely controlled by encapsulation and by resin polymer, waxes and sulfur coating	Excess release produces toxicity and destroy ecological balances
Effective duration of release	It extends the effective duration of supply of nutrients to plants	Nutrients used by plants at the time of delivery and others are lost as insoluble salts
Loss rate of fertilizer nutrients	It reduces losses of nutrients by leaching, runoff and drift	High loss rate of fertilizers by leaching, runoff and drift

Table 9.
Nano-fertilizers vs. conventonal fertilizers.

Encapsulated nano-silica, nano-additives (TiO_2), PMAA, and Chitosan PMAA (addition of 400 ppm in fertilizers).

8.2.2 Biofertilizers

Drawbacks (a) short shelf life, (b) temperature sensitivity and (c) storage & desiccation problem.

Solution: Polymeric nano-particle for coating (resistance against desiccation), Water in oil emulsion technique, Hydrophobic silica nano-particle addition.

Gold nano-particles increase the growth of *Pseudomonas flurescence*, *Pseudomonas elgii*, and *Bacillus subtilis* which are considered nano-bio fertilizers.

9. Commercial products from three categories of nano-fertilizers

i. Nanogro: (Fe + Co + Mg + Mn mixed in pharmaceutical sugar): homeopathic plant medicine (**Table 10**) [12].

ii. Silicon23 + Microbes N,P,K: plant growth regulator

iii. Lithovit high yield fertilizer: increase CO_2 content in foliage to increase photosynthesis.

iv. NanoNB: fertilizer product from the UK: blend of humified organic material + N + B; acts as a plant growth stimulator.

v. NanoGreen: alkylamine + non-ionicsurfactant + organicalkohol: accelerate the rate of photosynthesis by entering the nano-molecule through stomata.

vi. Geohumus: water retaining hybrid materials (inorganic-organic-polymeric molecule).

10. Beneficial and detrimental effect of nano-particles

- Environmental accumulation in soil and water and due to their small size, it is likely they will become airborne.
- Edible plants have been shown to uptake and accumulate nano-particles.

N nano fertilizers	P nano fertilizers	K nano fertilizers	Micro nutientnano fertilizer
Poly olefin resin coated urea	P- rock + NH_4^+ Zeolite	K- Zeolite	Zn Zeolite, ZnO
Neem coated urea	Clinoptilonite Zeolite	Surface modified Zeolite	Cu-Zeolite, CuO
Nano Sulfur coated urea	Zeolite + Ca-P mineral apatite	Bentonite	MCM-41 for Mo
Surfacte modified Zeolite	Nano apatite	Chabazite	Fe_3O_4 and FeO

Field application of nano-fertilizers: nano-fertilizers has the potential contribution in slow release of fertilizers.

Table 10.
Sources of various kinds of nano-fertilizers (NPK and micro-nutrient based nano fertilizers).

- The use of nano-particles can adversely affect soil microbiota, creating an imbalance in bacterial diversity.
- Nano-particles with different compositions are associated with health effects in humans and animals, such as arteriosclerosis, high blood pressure, blood clots, stroke, arrhythmia, heart disease, heart attack, respiratory diseases, neurodegenerative diseases, reproductive system diseases, and various cancers.
- Nano-particles will become ubiquitous in the soil, atmosphere, and water and will be available for uptake in other plant species for which they are phytotoxic.
- Nano-particles affect more dangerously in the F2 generation of the crop.

11. Limitation for the adoption of nano-fertilizers in agriculture

- Relatively slow progress in fertilizer formulations
- Lack of clarity on regulations and innovations in fertilizer industries
- Increase in cost of fertilizers due to the use of design polymers as nano-coatings
- Lack of overall standardization in the fields
- Most of the formulations are claimed as nano, but those are in micron or submicron levels
- Need to incorporation of molecular recognition agents such as antibodies to aid in specificity in fertilizer nutrients.

12. Conclusion

- The use of nano-materials as a fertilizer delivery is expected to reduce the doses and ensure control release.
- The use of nano-particles as a fertilizer in various cereal crop provide higher yield and productivity.
- Nano fertilizer improves the biological yield of pulses and *Brassicaceae crops.*
- Nano-fertilizers in excess amounts cause harmful effects in plants and enters the food chain.
- Nano-fertilizers' uses have beneficial and detrimental effects on soil, plants, animals, and the environment.

13. Future prospects for the adoption of nano fertilizers in agriculture

- The full potential of nano-fertilizers is yet to be realized.
- Plant-nanoparticle-soil interaction need to be further understood.

- Effect of nano-fertilizers on the environment and human health also needs to be investigated in detail.
- Lab to the land concept: uses of nano-fertilizers still lack in field application on a comprehensive basis. It is still using pot culture studies, but it can be used in the field condition.
- Economics of nano-fertilizers: as it is used in low doses and losses of nutrients due to application of nano-fertilizer is minimum, it can be treated from an economic point of view.
- Production on a commercial scale: there is an urgent need to produce nano-fertilizer on a commercial scale for farmer use.

Author details

Rahul Sadhukhan[1,2*], L. Devarishi Sharma[2,3], Suman Sen[4], Snehashis Karmakar[5], Koushik Banerjee[6] and Kirtiranjan Baral[4]

1 Agronomy, MTTC and VTC, Mizoram, India

2 CAU, Imphal, India

3 Soil Science and Agricultural Chemistry, MTTC and VTC, Mizoram, India

4 Division of Agronomy, ICAR-IARI, New Delhi, India

5 Department of Plant Physiology, BCKV, Mohanpur West Bengal, India

6 Division of Agricultural Physics, ICAR-IARI, New Delhi, India

*Address all correspondence to: rahulsadhukhan92@gmail.com

References

[1] Subramanian KS, Manikandan A, Thirunavukkarasu M, Rahale CS. Nano-fertilizers for balanced crop nutrition. In: Nanotechnologies in Food and Agriculture. Cham: Springer; 2015. pp. 69-80

[2] Dey JK, Das S, Mawlong LG. Nanotechnology and its importance in micronutrient fertilization. International Journal of Current Microbiology and Applied Sciences. 2018;7(5):2306-2325

[3] Ali ASM, Sato Y, Shimizu N, Ichinose T. Exploration and analysis of socioeconomic potentials for sustainable development in rural Egypt using GIS technology. Journal of Rural Planning Association. 2001;**20**(20):175-180. DOI: 10.2750/arp.20.20-suppl_175

[4] Seleiman MF, Almutairi KF, Alotaibi M, Shami A, Alhammad BA, Battaglia ML. Nano-fertilization as an emerging fertilization technique: Why can modern agriculture benefit from its use? Plants. 2021;**10**(1):2

[5] Fleischer MA, O'Neill and Ehwald, R. The pore size of non-graminaceous plant cell wall is rapidly decreased by borate estercross-linking of the pectic polysaccharide rhamnogalacturon II. Journal of Plant Physiology. 1999;**121**: 829-838

[6] Manjunatha SB, Biradar DP, Aladakatti YR. Nanotechnology and its applications in agriculture: A review. Journal of Farm Science. 2016;**29**(1): 1-13

[7] Subramanian KS, Rahale CS. Ball milled nanosized zeolite loaded with zinc sulfate: A putative slow release Zn fertilizer. International Journal of Innovative Horticulture. 2012;**1**(1):33-40

[8] Bakhtiari M, Moaveni P, Sani B. The effect of iron nanoparticles spraying time and concentration on wheat. Biological Forum—An International Journal. 2015;7(1):679-683

[9] Davarpanah S, Tehranifar A, Davarynejad G, Abadía J, Khorasani R. Effects of foliar applications of zinc and boron nano-fertilizers on pomegranate (*Punicagranatum*cv.*Ardestani*) fruit yield and quality. Scientiahorticulturae. 2016;**210**:57-64

[10] Tarafdar JC, Raliya R, Mahawar H, Rathore T. Development of zinc nanofertilizer to enhance crop production in pearl millet (*Pennisetumamericanum*). Agricultural Research. 2014;**3**(3):257-262

[11] Malandrakis AA, Kavroulakis N, Chrysikopoulos CV. Use of copper, silver and zinc nanoparticles against foliar and soil-borne plant pathogens. Science of the Total Environment. 2019; **670**:292-299

[12] Mastronardi E, Tsae P, Zhang X, Monreal C, DeRosa MC. Strategic role of nanotechnology in fertilizers: Potential and limitations. In: Nanotechnologies in Food and Agriculture. Cham: Springer; 2015. pp. 25-67

Chapter 3

Nano Pesticides Application in Agriculture and Their Impact on Environment

Malik Asif, Shayesta Islam, Mushtaq A. Malik, Zaffar Mahdi Dar, Amjad Masood, Saima Shafi, Bisma Rashid and Showkat Sidique

Abstract

Environmental contamination and the tolerance developed by the pests, pathogens are some of the environmental issues related to the aimless utilization of chemical pesticides. It has became matter of serious concern for environment, food quality and soil health. Nanotechnology, envisaged as a swiftly emerging field has capability to reform food systems in agriculture. Nanotechnology provides an imperishable solution to these problems by the establishment of nano-pesticides. The functional components or the conveyor molecules used are of nano size. The performance of these nano sized particles is much better the traditional pesticides, as the smaller size aids in proper spreading on the pest surface. Amelioration in solubility of operational components, betterment in stability of formulation, gradual liberation of operational components and enhancement in mobility are some of the paramount advantages of nano particles attributed to the minute size of particles and greater surface area. Thus, nano particles have strengthened activity against target pests in comparison to bulk materials. Furthermore, nano-formulations sustain productive use in agriculture by offering systemic properties, uniform leaf coverage and enhanced soil properties. Despite all the positive aspects, it might have certain negative effects as well, like exposure of humans through distinct routes Viz, exposure to nano pesticides either directly or indirectly like adsorption through skin, or inhalation while breathing air or transfer from one energy level to another by taking contaminated food and water.

Keywords: nanopesticide, formulations, environmental risk, agriculture, future aspect

1. Introduction

Pest management is still greatly dependent on the utilization of pesticides, through the application of organic chemical originated components on crops, products and urban habitat, regardless of divergent alternative methods available. Most of the registered pesticides intrude with the nervous system of insects through their prime method of action, that is they are neurotoxic, and are considered to enhance the threat of expansion of various neurodegenerative diseases, such as

Parkinson's disease [1]. In order to lessen the utilization of neurotoxic compounds, various novel products have been launched during recent times in the market, which include insect killing agents like, destroying agents of adenosine triphosphate (ATP) or controlling agents of insect growth, in spite of this environmental affect is a matter of great concern. In this connection, the matter has been correlated to the lethality caused to mammals, environmental pollution, and biomagnifications in the food chain. Along with amplified tolerance rate developed by insects species to currently used products, the aforesaid concerns have became fundamental challenges in agriculture, and may significantly restrict the beneficial active compounds. To tackle this trouble, there is need of novel pest control strategies for pest management through the initiation of novel ideas related to pest tolerance and advancement in technologies. Nanotechnology has come up with as considerably alluring field of research to achieve these aims, by furnishing advanced methodologies for designing novel operational components with microscopic measurements, formation procedures and dissemination, which are collectively referred to as "nanopesticides." Nanopesticide research, is the introduction of nanotechnology to protect crops. This field encompasses extensive research aspects viz.,

- Learning the basic interrelation notion of nano-scale materials.
- Preparation of nanoemulsions from active components and their dispersal using pesticides.
- Utilizing nanomaterials as active agents to develop novel nanopesticide formulations.
- Using these nanomaterials as nanocarriers for pesticide delivery [2–4].

This nanopesticide research is comprehensive and is believed to address the major drawbacks of the existing pest management strategies and comes up with new advanced nano-based formulations that stay stable and active, that is not influenced by sun, heat, and rain in the target environment, enter the target organism (insect), resist defense of the pest, remain amiable to plants and mammals, formulation and manufacture is lucrative, and ideally possess a new-fangled approach of action [2, 5, 6]. NPs in diverse fashion and chemical make-up including metal, metal oxides, semiconductor quantum dots (QDs), carbon, ceramics, silicates, lipids, polymers, proteins, dendrimers, and emulsions were made by utilizing a wide range of materials synthesized or used from natural materials [7, 8]. Some familiar advantages of NP-based pesticide formulations include:

- Solubility of water-insoluble operative constituents is enhanced.
- Stability of the formulation is elevated.
- Eradicates dangerous organic solvents.
- Ability to liberate operative constituents slowly.
- Early deterioration is averted by improvement in stability.
- High mobility and insecticidal activity is attributed to nano size
- Endurance is believed to get increased due to large surface area [9].

The ideas related to advancement in the manufacturing of nano based formulations are

1. The advancement in traditional pesticide formulations.

2. The establishment of delivery systems.

3. Utility of NPs as nano carriers, and use of solid NPs as operative agents.

2. Nano pesticide formulations

Utilization of technologies such as encapsulation and restrained liberation methods for the use of pesticides is of remarkable concern. Various researchers have laid emphasis on the importance of Nanocapsules and nanoparticles for agricultural purposes [10]. A number of companies have prepared different formulas that comprise of NPs with the dimensions ranging from 100 to 250 nm. A few make use of Nanoemulsions, which are prepared by dissolving nanoparticles pesticides evenly either in water or oil with dimensions ranging from 200 to 400. The manifold applications can be achieved like protective measures, treatment, or preventing reaped product by easily integrating these emulsions into gels, creams, liquids. One of the modern desired restricted liberation of agricultural chemicals is the utilization of materials incorporated with silica. Porous hollow silica NPs (PHSN) was used as pesticide carrier to record the restricted discharge pattern of pesticide namely avermectin [11]. It was revealed that PHSNs can be utilized to restricting pesticide dissemination applications, as these carriers noticeably deferred the release of the pesticide. NPs can imbibe and bind with other compounds effortlessly because of large surface areas, flow without difficulty in lepidopteran systems and certainly be utilized for pesticide development [12]. It has been reported that most terpene compounds have antifeedant action and are extremely volatile. Insecticidal activity and life span of the formulations significantly enhanced by integrating certain plant extracts with nano-silica [13]. Formulations made by integrating a-pinene and linalool with nano-silica improved biological activity of the plant pure chemicals, stability of the formation with superior zeta potential, restricted liberation of the plant derived components compound, and prolonged life span of the plant isolates [13]. These formulations exhibit great antifeedant action against *Spodoptera litura* and *Achaea Janata* (L. These nanoformulations smoothly disseminate, which was made clear by the dissemination dispersion studies. It was concluded that life span analysis of nano formulations with terpenes have no impact on dissemination, dimensions, zeta potential, or biological activity of the nanoformulations up to period of 6 months after harvesting [3, 13].

3. Solid nanoparticles as nanopesticides

A number of nanoparticles have been recommended as potential candidates both as nano carriers as well as operational pesticide agencies or biological pesticides, because of intrinsic antibacterial/pesticidal properties.

3.1 Inert dusts as insecticides

The mechanism of activity of stable dusts such as silica, alumina, and clay is due to the destruction of the waxy coating on the insect's epidermis by surface assimilation and scraping. This destruction trigger insects to dehydrate, dry out and die. Because the mechanism of action is physical, it is difficult to acquire insect resistance [14].

3.1.1 Silica

The utility of silica based nanoparticles (SNPs) for controlling pests in agricultural is reasonably recommended because silicon has the capacity to escalate the resistance of plants to abiotic and biological stresses [12]. The deadly effects of unstructured water loving, and oil loving SNPs to the rice weevil *Sitophilus oryzae* were studied by [15] and found that they were much more helpful and efficient than bulk silica. No new infections were detected after depositing for period of 60 days. Silica particles are physically absorbed by lenticular lipids that destroy the protective cuticle and destroy insects purely by physical means with a similar mode of action observed on diatoms particles used for protection of stored grain [12, 16]. In addition, modified hydrophobic surface-charged silica (3-5 nm) have been fruitfully used in veterinary medicine to control a wide range of essential agricultural and animal ectoparasites [17]. The fungal growth was decreased and germination of seeds was promoted by effective application of surface charged silica as a thin film of seeds [18]. The metabolic process of certain agricultural crop groups and horticultural plants does not get affected by the superficial application of NPs to the leaves and stem. They do not encourage any modification in gene expression in insect organs, so they deserve approval as nano pesticides. Additionally, World Health Organization (WHO) declared the utilization of unstructured silica as a nanopesticide risk free for human consumption. In traditional pesticides these SNPs also serve as carriers. Comparison to small amounts of chlorfenapyr, the biological effectiveness was enhanced by loading the insecticide chlorfenapyr into dispersed SNPs [19]. This is assigned to the nanoscale size and intrinsic insecticidal activity of SNPs. In addition, it has been reported that porous hollow SNPs can safeguard the fleeting insecticide avermectin from UV degradation and liberate it for a long time. SNP carriers have been observed to promote the sustained release of avermectin for approximately 30 days [20].

3.1.2 Diatomaceous earth (DE)

DE is comprised predominantly of unstructured silica originated by decomposing small sized plants under high temperature and pressure. These are employed to safeguard stored grain from pest insects. It has deleterious effects on grain properties especially on bulk density because it is required in high doses. Researchers are engaged to integrate DEs with other insecticides to attain high efficiency, so as to make it effective in low doses. For example, combination of DE and plant decoction bitterbarkomycin was used to control grain pest *Rhyzopertha dominica*, and was recorded to be efficient at low dose concentration of 150 ppm [21].

3.2 Alumina

Nanostructured alumina (NSA) was recognized as a successful insecticide in case of grain pests, *S. oryzae* (L) and *R. dominica* (F) by [22]. They also collated NSA formulations against a foremost traditional DE commodity and declared their formulation to be efficient. NSA dusts manufactured using a modified glycine-nitrate combustion process were reported to be more effective against, *S. oryzae* (L) and *R. dominica* (F) in the recent research. The results revealed that, key factors determining insecticidal efficacy are dimension of particle, surface area, and external features. However, these are not believed to be only factors responsible, and to attain superior outcomes for specific species, there is need to improve preparation pathways [23].

3.3 Clays

Nanoclays are fine layers of silicate minerals with following dimensions (thickness 1 nm and width 70–150 nm). Source of nanoclays is montmorillonite clays, which are mostly found in volcanic ash, formed by size reduction and surface modification and are biocompatible and have less hazardous associated. Among these living materials anionic clays are found to be most promising. They have fruitfully served as conveyor for the α-naphthalene acetate, which is regulator for growth in plants and for the restrained liberation of the 2,4-dichlorophenoxy-acetate. Natural antibiotic cinnamate, which is utilized for pest management, is susceptible to quick deterioration in soil and is required in significant concentration. It was declared that by loading it on double layered hydroxide low pace liberation of the antibiotic and prolonged longevity in the soil is achieved [24]. This indicates the outstanding potential of nano-clays to be used for low pace/target specific dissemination of pesticides and fungicides.

4. Preparation of nano pesticides by encapsulating pesticide nano particles

Encapsulation is a process of encompassing or enveloping one substance inside another fabric on a very little scale or may be characterized as the method by which a material is encompassed and safe guarded from external conditions that would break it down, yielding capsules extending from less than one micrometer to several hundred micrometers in size [25]. Encapsulation can be obtained by few strategies with distinctive purposes in mind. Encapsulation of materials may take place deliberately that the central material be restricted inside capsule walls for a particular period of time. On the other hand, central substances can be incorporated so that the core material will be released either progressively through the capsule walls, known as controlled release or diffusion, or when outside conditions activate the capsule walls to break, melt, or dissolve [26]. The different areas in which encapsulated materials have numerous applications include agriculture, pharmaceuticals, foods, cosmetics and fragrances, textiles, paper, paints, coatings and adhesives, printing applications, and many other industries. Encapsulation permits the farmers to apply pesticides less frequently instead of requiring profound concentration and toxic initial applications to be released over time, allowing farmers to apply the pesticides less often rather than requiring very highly concentrated and perhaps toxic initial applications followed by repeated applications to fight the losses due to leaching, evaporation, and degradation [27]. Protection of pesticides from full contact with the components reduces the risk to the environment and those that might be exposed to the chemicals and provides a more efficient technique for pest control [28].

5. Toxicity and environmental impacts

The vulnerability and environmental consequences of nanoscale-based formulations is of great concern, which needs to be addressed. The amount of nanoformulations in soil, surface water, and groundwater, and effects impacts on non-target organisms have not been predicted much. The fate of nano formulations depends on number of chemical parameters, like pH, ionic strength, and dissolved molecules in the media [29]. Emphasis is laid on certain fundamental concerns mentioned under following sections.

5.1 Direct toxicity on humans and environment

The harmfulness of numerous pesticides utilized in a nano-scale formulation has been determined by specialists, the probable negative impact and the probable impacts of nano dimensions must be taken in to consideration during nano-scale formulations. More inquiry is needed to affirm the harmfulness of nano-scale formulations and the factors contributing to harmfulness of nano-particles like size, charge, shape, and chemistry.

5.2 Durability and persistence

The fundamental criteria for analyzing dangers associated with the application of nano pesticides is their longevity in the environment. The non target organisms get exposed when there is suspended liberation or release for a prolonged period of time [30].

5.3 Bioavailability

The nanocarriers promote transfer of some of the immobilized operational components, which is considered one of the likely detrimental effects. Thus increase the availability to organisms [31]. Enhancement in absorption by target organisms have been reported by certain nanoformulations. It must be ensured that no damage caused to non-target organisms.

5.4 Release profile/degradation

The two important factors which govern the liberation of operational components involve class of nanocarrier and the dissemination of active components in matrix. Research suggested that environmental affect and virulence of operational ingredients is contributed by set of factors viz., surface desorption, dissemination via the polymer matrix, and deterioration of the polymer [31].

5.5 Fate of the carrier

The centre of attention of research now are carriers established using natural biologically degradable polymers like, polysaccharides, or lipids, which deteriorate into by-products of least concern [13].

6. Nanopesticides interaction with other environmental contaminants

The evaluation of pesticides is done on a product-by-product basis, in a similar fashion to other contaminants. In the real environment, mixed interactions are possible as they co-occur with other substances. A nanopesticide may have collaborative interaction with other contaminants through the Trojan-horse effect, so hazard analysis of environmental mixtures is generally not needed for product hazard estimation [24]. In this effect the substance are carried to a tissue of organism or to an organism by interaction of substance with ENP, due to which the interior exposure to contaminants is enhanced, and might not get accumulated. The need of hour is to integrate these interrelations into authoritative hazard analyzing schemes.

7. Nano-pesticides for greener agriculture

Nano-pesticides are defined as preparations that deliberately introduce elements in the nm size range and the characteristics related with this small size range are new, these nanopesticides have earlier been launched in the market. Nanopesticides cannot be treated as a single category but have been blended in a numerous products. The various classifications of Nanopesticides include organic components viz., A.I., polymers and inorganic components like oxides of metals in diverse configurations like particles and micelles [32]. One of the most economical and multifaceted method of containing insect pests is the residues remaining on the surface after application. Microencapsulation; a nanotechnological approach can be employed to modify the insecticidal value by protecting the operational components from environmental conditions and by promoting persistence. The usage of engineered nanoparticles (ENPs) is gaining considerable attention in the pesticide sector due to the establishment of a wide varieties of plant protection products termed as "nanopesticides". Nanopesticides "involve either very small particles of a pesticide active ingredient (A.I) or other small engineered structures with useful pesticidal properties". The advantages offered by this emerging technology involves amplified efficacy, durability, and a cutting down the amount of active ingredient required. Nanoemulsions, nanocapsules like polymer and commodities containing pristine engineered nanoparticles, such as metals and their oxides, and nano-clays have been recommended. These products can be exploited to intensify the efficiency of prevailing pesticide operational components or results in improvement of environmental safety profiles or both [30].

8. Nano pesticide: future possibilities

Recent investigations have shown that nano-pesticides can reduce the deleterious effects of chemical based pesticides and furnish target-specific control of pests, and help develop intelligent nano-systems for minimizing problems like environmental imbalance, and negative effects on food security, and crop productivity [33]. They are effective for long term utility and provide solution to environmental related problems like nutrient richness in water bodies and accumulation of non-biodegradable components in the food chain due to restricted liberation of operational ingredients. Furthermore, nano-pesticides show efficient pest control property due to amplified solubility and stabilities of operational components [34]. Still, there is necessity to modify the techniques in order to have remarkable benefits to agriculture. A few aspects for nano-scale pesticide delivery platforms discussed by [33] in their review include:

- The effectiveness of nanopesticide development is enhanced by use of green chemistry and environmentally viable principles [35].
- Upgrading traditional utilization of nanopesticide.
- Realistic utilization should be estimated at field level by comparing it with traditionally used products.

- The vulnerability of nanopesticides is estimated by Environmental impact assessment.
- Modification in the policy for application of nano material in agriculture.
- Agrochemical industries will come up with many solutions by the launching of smart nanopesticide, that is solubility of the operational components, stability, restricted liberation, and dissemination of active components on specific organisms, but lot of research need to be done to get acquainted with the fate of nanopesticide in the environment.

Author details

Malik Asif[1*], Shayesta Islam[2], Mushtaq A. Malik[1], Zaffar Mahdi Dar[1], Amjad Masood[3], Saima Shafi[1], Bisma Rashid[1] and Showkat Sidique[4]

1 FOA-Wadura, Division of Basic Sciences and Humanities, SKUAST-Kashmir, India

2 FOH-Shalimar, Division of Environmental Sciences, SKUAST-Kashmir, India

3 FOA-Wadura, Division of Agronomy, SKUAST-Kashmir, India

4 FOA, Wadura, Division of Agriculture Statistics and Economics, SKUAST-Kashmir, India

*Address all correspondence to: drasif_skuast@yahoo.com

References

[1] Ascherio A, Chen H, Weisskopf MG, Reilly EO, McCullough ML, Calle EE, et al. Pesticide exposure and risk of Parkinson's disease. Annals of Neurology. 2006;**60**:197-203

[2] Smith K, Evans DA, El-Hiti GA. Role of modern chemistry in sustainable arable crop protection. Philosophical Transactions of the Royal Society. 2008;**363**:623-637

[3] Usha RP, Madhusudhanamurthy J, Sreedhar B. Dynamic adsorption of a-pinene and linalool on silica nanoparticles for enhanced antifeedant activity against agricultural pests. Journal of Pest Science. 2014;**87**:191-200

[4] Benelli G, Lukehart CM. Special issue: Applications of greensynthesized nanoparticles in pharmacology, parasitology and entomology. Journal of Cluster Science. 2017;**28**(1):1-2

[5] Benelli G. Plant-mediated biosynthesis of nanoparticles as an emerging tool against mosquitoes of medical and veterinary importance: A review. Parasitology Research. 2016a;**115**(1):23-34

[6] Benelli G. Green synthesized nanoparticles in the fight against mosquito-borne diseases and cancer—a brief review. Enzyme and Microbial Technology. 2016b;**95**:58-68

[7] Niemeyer CM, Doz P. Nanoparticles, proteins, and nucleic acids: biotechnology meets materials science. Angewandte Chemie International Edition. 2001;**40**:4128-4158

[8] Oskam G. Metal oxide nanoparticles: Synthesis, characterization and application. Journal of Sol-Gel Science and Technology. 2006;7:161-164

[9] Sasson Y, Levy-Ruso G, Toledano O, Ishaaya I. Nanosuspensions: Emerging novel agrochemical formulations. In: Ishaaya I, Nauen R, Horowitz AR, editors. Insecticides Design Using Advanced Technologies. Springer: Berlin; 2007. pp. 1-32

[10] Pavel A, Creanga DE. Chromosomal aberrations in plants under magnetic fluid influence. Journal of Magnetism and Magnetic Materials. 2005;**289**: 469-472

[11] Wen LX, Li ZZ, Zou HK, Liu AQ, Chen JF. Controlled release of avermectin from porous hollow silica nanoparticles. Pest Management Science. 2005;**61**:583-590

[12] Barik TK, Sahu B, Swain V. Nanosilica—from medicine to pest control. Parasitology Research. 2008;**103**:253-258

[13] Madhusudhanamurthy J, Usha Rani P, Sambasiva RKRS. Organic–inorganic hybrids of nano silica and certain botanical compounds for their improved bioactivity against agricultural pests. Current Trends in Biotechnology and Pharmacy. 2013;7:615-624

[14] Shah MA, Khan AA. Use of diatomaceous earth for the management of stored-product pests. International Journal of Pest Management. 2014; **60**(2):100-113

[15] Debnath N, Das S, Seth D, Chandra R, Bhattacharya S, Goswami A. Entomotoxic effect of silica nanoparticles against Sitophilus oryzae (L). Journal of Pest Science. 2011;**84**:99-105

[16] Korunić Z. Diatomaceous earth, a group of natural insecticide. Journal of Stored Products Research. 1998;**34**:87-97. DOI: 10.1016/S0022-474X(97)00039-8

[17] Ulrichs C, Mewis I, Goswami A. Crop Diversification Aiming Nutritional

Security in West Bengal: Biotechnology of Stinging Capsules in Nature's Water-Blooms. Ann Arbor: Ann Tech Issue of State Agri Technologists Service Assoc; 2005. pp. 1-18

[18] Robinson D, Salejova-Zadrazilova G. Observatory nano. In: Nano Tachnologies for Nutrient and Biocide Delivery in Agricultural Production. Working Paper Version. 2010

[19] Song MR, Cui SM, Gao F, Liu YR, Fan CL. Dispersible silica nanoparticles as carrier for enhanced bioactivity of chlorfenapyr. Journal of Pesticide Science. 2012;**37**(3):258-260

[20] Li ZZ, Chen JF, Liu F, Liu AQ, Wang Q, Sun HY, et al. Study of UV-shielding properties of novel porous hollow silica nanoparticle carriers for avermectin. Pest Management Science. 2007;**63**(3):241-246

[21] Athanassiou CG, Korunic Z. Evaluation of two new diatomaceous earth formulations enhanced with abamectin and bitterbarkomycin, against four stored-grain beetle species. Journal of Stored Products Research. 2007;**43**(4):468-473

[22] Stadler T, Buteler M, Weaver DK, Sofie S. Comparative toxicity of nano structured alumina and a commercial inert dust for Sitophilus oryzae (L.) and Rhyzopertha dominica (F.) at varying ambient humidity levels. Journal of Stored Products Research. 2012;**48**:81-90

[23] Buteler M, Sofie SW, Weaver DK, Driscoll D, Muretta J, Stadler T. Development of nanoalumina dust as insecticide against Sitophilus oryzae and Rhyzopertha dominica. International Journal of Pest Management. 2015;**61**(1):80-89

[24] Park EJ, Yi J, Kim Y, Choi K, Park K. Silver nanoparticles induce cytotoxicity by a Trojan-horse type mechanism. Toxicology In Vitro. 2010;**24**(3):872-878

[25] Del Carmen Gimenez-López M, Moro F, Torre AL, Gómez-García CJ, Brown PD, Slageren JV, et al. Encapsulation of single-molecule magnets in carbon nanotubes. Nature Communications. 2011;**2**:407

[26] Doane WM. Encapsulation of pesticides in starch by extrusion. Industrial Crops and Products. 1992;**1**:83-87

[27] Lewis DH, Cowsar DR. Principles of controlled release pesticides. In: Scher HB, editor. Controlled Release Pesticides. Washington, DC: American Chemical Society; 1977. pp. 1-16

[28] Ulbricht H, Hertel T. Dynamics of C60 encapsulation into single-wall carbon nanotubes. The Journal of Physical Chemistry. B. 2003;**107**:14185-14190

[29] Kookana RS, Boxall ABA, Reeves PT, Ashauer R, Beulke S, Chaudary Q, et al. Nanopesticides: Guiding principles for regulatory evaluation of environmental risks. Journal of Agricultural and Food Chemistry. 2014;**62**:4227-4240

[30] Kah M, Beulke S, Tiede K, Hofmann T. Nano-pesticides: State of knowledge, environmental fate and exposure modelling. Critical Reviews in Environmental Science and Technology. 2013;**43**(16):1823-1867

[31] Kah M, Hoffmann T. Nano pesticide research. Current trends and future priorities. Environment International. 2014;**63**:224-235

[32] Ragaei M, Sabry AH. Nanotechnology for insect pest control. International Journal of Science, Environment and Technology. 2014;**3**:528-545

[33] Nuruzzaman M, Rahman MM, Liu Y, Naidu R. Nanoencapsulation, nano-guard for pesticides: A new

window for safe application. Journal of Agricultural and Food Chemistry. 2016;**64**(7):1447-1483

[34] Venugopal NVS, Sainadh NVS. Novel Polymeric Nanoformulation of Mancozeb – An Eco-Friendly Nanomaterial. International Journal of Nanoscience [Internet]. World Scientific Pub Co Pte Lt; Aug 2016;**15**(04):1650016. Available from: http://dx.doi.org/10.1142/s0219581x16500162

[35] Ocsoy I, Paret ML, Ocsoy MA, Kunwar S, Chen T. Nanotechnology in plant disease management: DNA-directed silver nanoparticles on graphene oxide as an antibacterial against Xanthomonasperforans. ACS Nano. 2013;**7**(10):8972-8980

Chapter 4

Vital Role of IPFT in Development of New-Generation Pesticide Formulation for Crop Protection: Advancement Overview in Asian Countries

Nusrat Iqbal, Amrish Agrawal, Md. Imteyaz Alam and Jitendra Kumar

Abstract

The agricultural sector of Asian countries supports 60% of the global population, accounting one-fifth of the world's agricultural land. Despite the gap between demand and supply of food is gradually increasing due to the damages caused by insect and other pest attacks on the limited agricultural land, the pest attack has influenced the entire agriculture sector either directly or indirectly, causing socioeconomic losses. To combat, farmers have been using conventional agrochemicals nonjudiciously that lead to adverse effects such as pesticide resistance, environmental contamination, and non-target toxicity. In this regard, new-generation agrochemical formulation techniques are advantageous over conventional pesticides and play a vital role in sustainable agriculture by fulfilling the demand of over-rising food supply to feed the increasing population. These formulations exhibit desired bio-efficacy at lower doses and have minimum possibility to leave pesticide residues in crop products and the environment. Institute of Pesticide Formulation Technology (IPFT), Gurugram, is one of the leading institutes in Asia, which is actively engaged in developing new-generation formulations to deliver safer, efficient, and environment-friendly pesticide formulations. So far, IPFT has developed 60 pesticide formulations and transferred technologies to different agrochemical industries globally. The new-generation formulations developed by IPFT mainly include microemulsion, nanoemulsion, capsulated suspension, nano-encapsulation, an emulsion in water, mixed formulations including several botanical pesticide formulations. The new advancement in pesticide delivery systems is very supportive in combating the crisis faced by the agricultural sector. In this chapter, formulation of different new-generation pesticides and their advancement are summarized.

Keywords: agriculture, agrochemicals, pesticides, new-generation formulations, pest, environment protection, user-friendly, crop protection

1. Introduction

Pesticides are the integral and significant part of modern agriculture. Pesticide usage is incredibly valuable for increasing food production in fulfilling the demand of over-rising population. It has been previously stated by Webster et al. [1] that without pesticide use considerable economic losses will occur. The application rate of agricultural pesticides has swiftly augmented in different developing and developed countries [2]. Therefore, pesticides help in significant increase of yield and prevent the crop losses due to pest and pathogen attack. Application in a smaller quantity gives positive results but large quantity causes negative effect on environment, human health, and other beneficial organisms such as insects responsible for pollination [3]. However, non-judicious use causes various detrimental effects on environment and damages the ecosystem including economic loss of the farmers.

In formulated forms, pesticides are effective in very small quantity. There are many conventional pesticide formulations available in the market, which are being marketed and used in Asian countries in the form of dustable powder (DP), wettable powder (WP), emulsifiable concentrate (EC), soluble liquids (SL), etc. These conventional formulations are effective but due to certain limitations such as toxicity, cost, pest resistance, environmental contamination, death of beneficial organisms (e.g., honey bee), and human health problems they attracted interests toward advanced and safer pesticide formulation techniques [4]. Interestingly, in past 1990s, Asian countries explored certain new-generation formulation, which were solvent free, long lasting, safe and caused no effect on non-target organisms [5, 6].

With the advancement of new formulation technologies in Asian countries, many new-generation formulations have been developed to maintain a special place in international markets. In Asian countries, government supports the pesticide supply for the enhanced food production but its overuse (conventional formulations) has resulted in pesticide residue problem causing non-target and ecosystem toxicity.

In order to improve the safety of the user and ecosystem from the hazardous pesticides, Institute of Pesticide Formulation Technology (IPFT), India, has been engaged in developing new-generation formulations of synthetic as well as botanical pesticides. This chapter summarizes all the formulation advancement developed by IPFT for the safe and smart delivery of pesticide and reduces the adverse effects associated with conventional formulations.

2. Basic concept of formulation

A pesticide is the main active ingredient (actual chemical that controls pest population), which may be of synthetic chemical or botanical origin. Various chemical- and botanical-based pesticides have different physical and chemical characteristics such as solubility, viscosity, and physical state. Moreover, different pesticides act differently for various types of pest populations; for instance, some are effective for crawling insects, while the others are for flying insects. If a farmer uses active ingredient in pure form, he will face assured difficulties as follows:-

i. Active pesticide in unformulated form cannot spread evenly.

ii. In pure form, pesticides show phytotoxicity and toxicity toward non-targets. Pesticide in unformulated form environment-related hazards will enhance.

iii. After formulation development, application process is easy and convenient.

To resolve all these issues, formulation scientists formulate the active ingredient into different forms. As per the Knowles, "A pesticide formulation is a mixture of active ingredient with different inert materials to improve its stability during storage, easy handling, improves safety, application, or effectiveness towards pest population" [5].

3. Conventional formulations and their limitations

In Asia and Pacific regions, the most common formulations being used are dustable powders, emulsifiable concentrates, wettable powders, granules, soluble liquids, etc., for the control of insect pest and many pathogenic diseases [7, 8]. The most common conventional formulations are discussed below:

1. **Dust formulations**:-These formulations are for contact action pesticides. These formulations are ready to use and no need for dilutions before application. They contain very low active ingredients (usually 1–10%). The inert ingredient mainly contains talc, ash, silica, bentonite, etc. (**Figure 1**). These are always applied as dry and can drift easily to non-targets. These formulations are mainly used for cracks and crevices.

2. **Granules (GR):** Granular pesticide formulation mesh size is higher as compared with dust. As per the British standard, mesh size is (250–1050 microns). For sufficient activity, at least 90% granules should be in the range of British standards. Due to larger size, drift velocity decreased and less wastage occurs over due to dust formulations [9]. In granular formulation, pesticide is encrusted onto or wrapped up in absorptive particles of silica, sand, clay, and shells of walnut or corn cobs pieces, etc., as carrier materials. Granular formulations are most suitable for pre-emergence herbicides and soil insecticides (**Figure 2**).

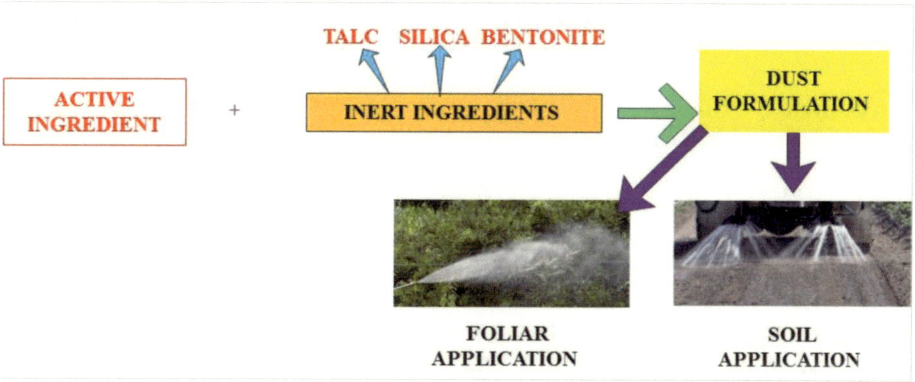

Figure 1.
Schematic illustration showing key steps involved in dust formulation and its mode of application.

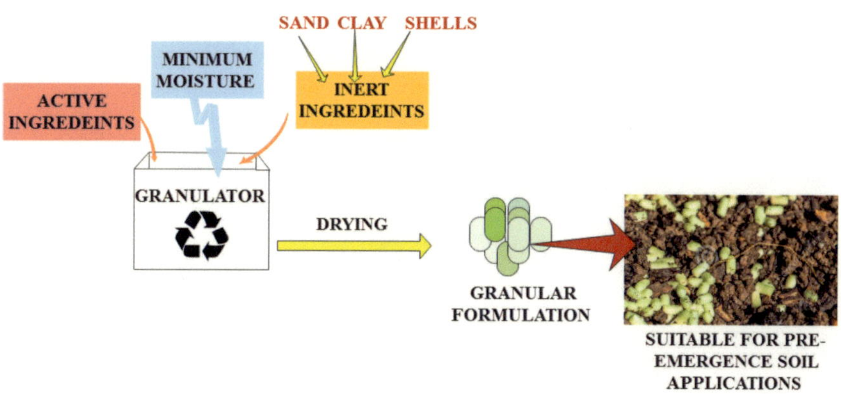

Figure 2.
Schematic representation showing key steps involved in granular formulation and its application.

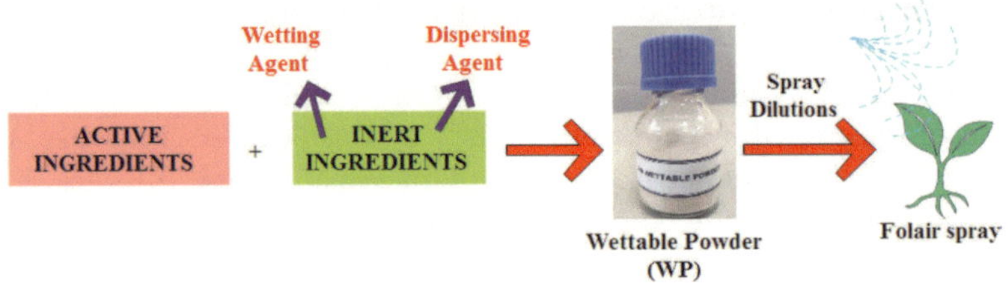

Figure 3.
Diagrammatical representation showing steps involved in WP formulation and its application.

3. **Wettable powder (WP):**-A WP contains active ingredient (pesticide) in a finely ground form along with wetting agents or dispersing agents (**Figure 3**). Wettable powders are applied by sprayers after diluting in water in the form of dilute suspensions. Wettable powders are safe to use, storing, and transporting.

4. **Soluble powders (SP):** In SP formulation, active ingredient or pesticide is mixed with the inert ingredients, which enhances the solubility of pesticide after dissolution in water (**Figure 4**). These SP formulations are easy to produce, economical, and stable under various temperature conditions.

5. **Emulsifiable concentrates:** Emulsifiable concentrate formulations are most common and popular formulation in Asian countries. These formulations are

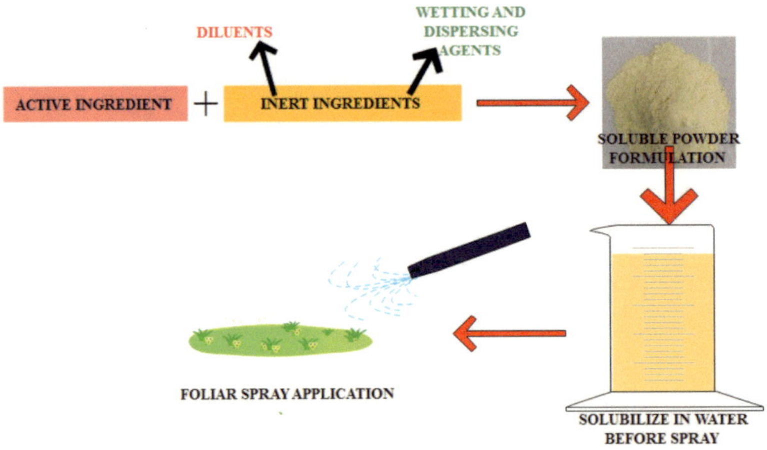

Figure 4.
Schematic representation showing key steps involved in soluble powder formulation and its application.

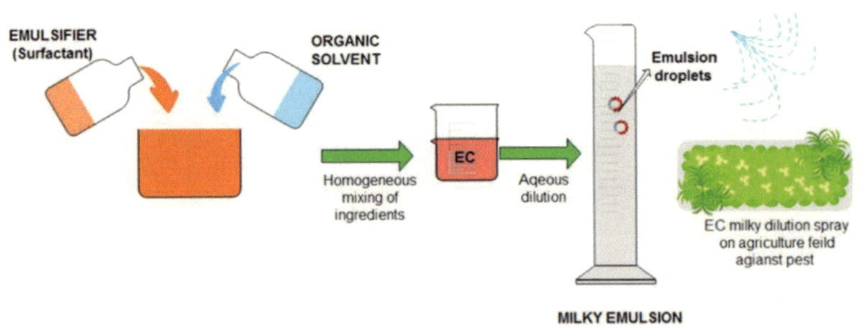

Figure 5.
Schematic representation showing key steps involved in EC formulation and its application.

suitable for the low melting point active ingredients, which are highly soluble in organic solvents. These formulations are developed by dissolving active ingredients in organic solvents along with emulsifying surfactants (**Figure 5**).

The main limitation is linked with the organic solvents that are mainly petroleum solvents being used in Asian countries. These petroleum solvents make the formulation flammable and cause dermal toxicity to the user. To rectify the problem associated with these conventional formulations, new formulations have been developed, which are improved and advance in terms of user and environment safety.

The major objectives of IPFT (India) to develop new formulation technologies are as follows:

1. To make the formulation techniques of different pesticide easy and convenient application.

2. To make pesticide formulations labor saving.

3. To prepare safer formulations for user and environment.

4. To reduce the toxicity of pesticides toward non-targets.

5. To minimize environmental contamination.

6. To enhance bio-efficacy against different types of pests.

7. To make the formulation economical and lower the frequency of applications.

4. Recent advancements in agrochemical formulations

Water-dispersible granules (WDG):-A WDG formulation is also termed as dry flowables (DF). These WDG formulations are non-dusty and disperse easily in water when added in spray tanks for its finer particle size in suspension. These are the safer and targeted delivery system for the various pesticides. Main uniqueness of these formulations is convenient application due to free-flowing nature and quick disintegration in water medium and applied as dilute suspensions. Size of the diluted suspended particles is very less, that is, 30–40 μm, therefore no nozzle clogging.

More advanced form of WDG is water-soluble bag-sealed WDG. In these packing bags, WDG are in partially disintegrated form, which can easily disintegrate in spray tanks (**Figure 6**).

Basic composition of WDG contains 50 to 90% pesticide (active ingredient) along with dispersing agents and wetting agents [10]. A dispersing agent is a type of surfactant that is added to formulations to improve the separation of the particles and inhibit particle size growth, and their settling or clumping. Pang et al. [11] reported lignosulfates as effective dispersing agent in water-dispersible granular formulations due to a high degree of sulfonation, high intrinsic viscosity property, and high molecular weights. Recently, sodium salt of methacrylic acid/styrene/sodium p-styrene sulfonate copolymer (SMSS) has synthesized a novel dispersant by free radical polymerization mechanism. This new dispersant has strong resistance to hard water and showed high performance and show above 90% suspensibility in hard water [12].

IPFT has developed water-dispersible granules of liquid pesticide such as triazophos. Recently, IPFT developed neem WDG for mosquito control. Other WG

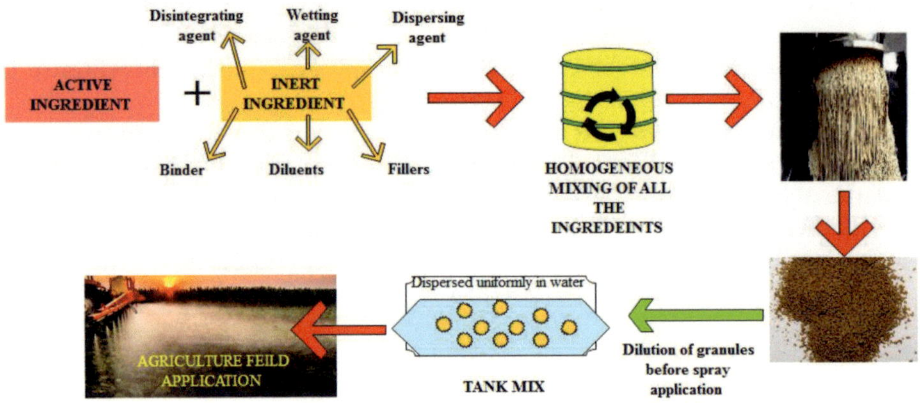

Figure 6.
Schematic representation showing key steps involved WDG formulation and its application.

formulations that have been developed by IPFT for agricultural usage are Captan 83WG, Isoproturon 75WG, Metamitron 70 WG, Mancozeb 75WG, Chlorothalonil 75WG, Endosulfan 75WG, Carbendazim 86WG, Divrinol 50WG, Thiram 80WG, Cypermethrin 40 WG, Thiamethoxam 25 WG, Deltamethrin 25 WG, and Triazophos 20 WG.

Suspension concentrate: Suspension concentrates are the stabilized dispersion of pesticides in water medium. This is the most popular formulation due to its safe and convenient use. The suspended particle size is very fine and provides good adhesion and penetration on target surface, which results in improved bio-efficacy. It has recently identified that SC formulations show less leaching of pesticides than conventional emulsifiable concentrate formulations. Similarly, another study has identified that in SC formulations, volatilization of active ingredient in environment reduced to 33.5% compared with conventional pesticide formulations [13]. Suspension concentrates have overcome the limitations associated with conventional organic solvent-based formulations (**Figure** 7).

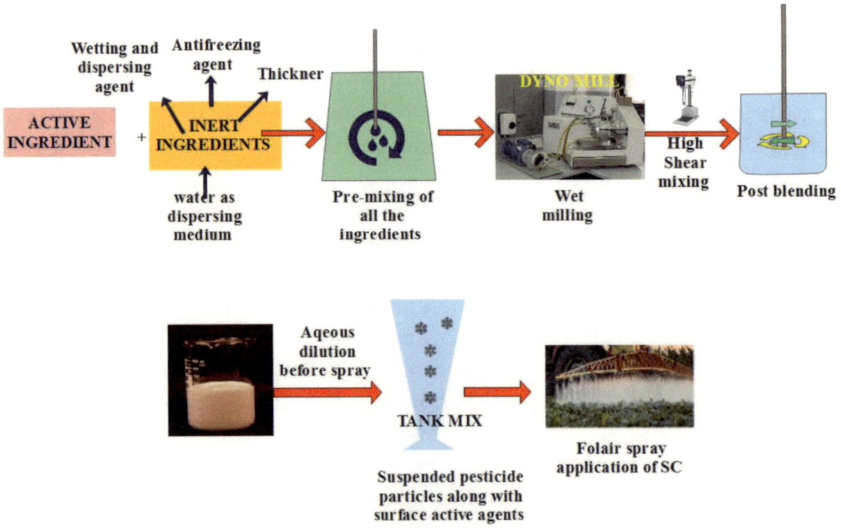

Figure 7.
Schematic key steps involved suspension concentrate formulation and its application.

The main advantages of suspension concentrates are easy application, no organic solvent used, and free from any toxicity to users and environment.

Careful selection of surfactant system is necessary to prevent hetero-flocculation (agglomeration of solid particles and droplets) during storage. There are three types of instabilities identified in SC during storage due to particle size change and they are particle aggregation, Ostwald ripening, and particle sedimentation [14].

5. Recent advances in suspension concentrate formulations

Effect of polymeric surfactant on physical stability has investigated in suspension concentrate. Polymeric surfactants provide high critical micellar concentration (CMC) and reduced Gibbs-free energy ΔG [15]. In another study, silicone surfactant is used as adjuvants in the SC formulation that results in high performance of surfactant and improves the physical stability and the quality of the formulation [16].

Particle size is also an important factor for stability as well as bio-efficacy enhancement. In a study, Vineela et al. [17] have reported that in *Bacillus thuringenesis* SC formulations, further reduction of particle size enhanced the bio-efficacy of formulation against *Spodoptera litura* (Lepidoptera: Noctuidae).

Some pesticides are photosensitive and easily decomposed. In a study, an amine-modified ligno-sulfonate surfactant was synthesized to make the SC formulation anti-photolysis [18]. Similar advancements have been conducted in IPFT on botanical-based formulations to inhibit photolysis. In addition to this, IPFT has developed several other UV protectant-based SC formulations for various SC formulations.

Entomopathogenic fungus, *Beauveria bassiana*, has been formulated as SC against *Helicoverpa armigera* larvae with LC_{50} value of 61.22 mg l^{-1} after 3 days of application [19]. Several other entomo-pathogenic fungi have been formulated as SC formulation by IPFT against various insect pests of different economically important crops such as spices and oil crops of Asian region.

Along with this IPFT has developed many synthetic and botanical pesticide SC formulations for agricultural pest control which include Isoproturon 50SC, Carbendazim 50SC, Sulfur 52SC, fipronil 5SC, thiomethxam 14.1% + Lamda cyhalothrin 10.6% SC, metamitron 70 SC, Neem SC, etc.

5.1 Capsule suspension (CS)

Capsule suspension (CS) is water-based slow release formulation containing active ingredient encapsulated inside microcapsules up to 10 microns size (**Figure** 8). CS formulation is a stable suspension of micro-capsules containing active ingredients and these microcapsules are synthesized by interfacial polymerization mechanism. CS formulations provide regulated, slow, and delayed the release of pesticides. This formulation is safer delivery mode by giving protection from toxic ingredients and stops the pesticide rate of degradation.

5.2 Advanced features of CS

CS are water-based (free from any organic solvent) homogeneous and uniform suspension formulation. Its application is safe, and provides enhanced and efficient bio-safety with reduced phytotoxicity. In order to develop CS formulation, the active ingredients should have low solubility in water and altogether hydrolytically stable. The pesticide from microcapsules gets slowly released after application. The CS formulations are slow release formulations and theses formulations prolong the availability of pesticide at target site. The CS formulation reduces environmental

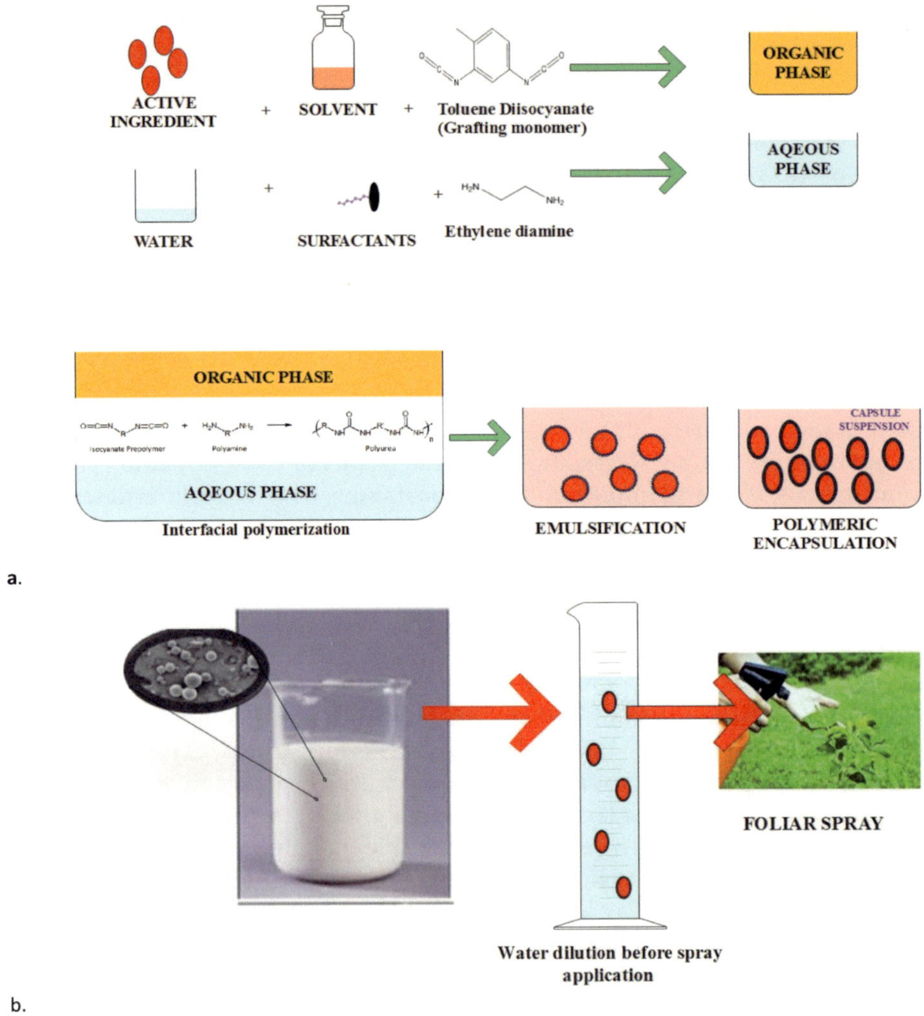

Figure 8.
Schematic representation of capsulated suspension formulation procedure a. and its application b.

contamination and leaching of pesticide and degradation of active ingredient by environmental factors like sunlight.

In a recent study, it was found that carboxy, methyl cellulose (CMC) and diallyl, dimethyl, ammonium chloride (DMDAAC) as monomers are the effective encapsulating agents of avermectin pesticide. This capsulated form gave higher performance and has excellent UV protecting property [20]. Thus, IPFT has developed CS for seed dressing for targeted delivery of pesticides without any wastage in environment [21].

Cyhalothrin is the most effective and broad-spectrum pesticide used in Central Asia. It has dual mode of action contact as well as systemic. Most efficient CS formulations developed by IPFT for sustainable agricultural applications are Lambda Cyhalothrion 10CS and Lambda Cyhalothrin 4.9CS.

5.3 Microemulsion

Microemulsion (ME) is the most efficient delivery system of botanicals as well as synthetic pesticides. By definition- "ME is a system of water, oil, and an amphiphile which is a single optically isotropic and thermodynamically stable liquid solution" [22].

Droplet size of ME is in nano-range, that is, 10 nm–20 nm. Based on dispersion medium, ME can be classified into two categories: O/W ME and W/O ME. Aqueous dilution of ME is required before spray application (**Figure 8**). Being a new-generation formulations, the characteristic features of MEs are nano sized, thermodynamically stable system with good penetrability, quick spreading ability, low and zero interfacial tension, and extended shelf life formulation.

6. Preparation methods of ME formulation

6.1 Phase inversion method

In phase dispersion method, dispersed phase is surfactant system along with active ingredient and water is the dispersant. The whole phase of inversion occurs in controlled temperature and other conditions for active kinetics (**Figure 9**).

Characteristic features of ME: ME formulation can effectively mask smell of the unpleasant active ingredients and protect pesticides from hydrolysis and oxidation. It enhances the solubility of water insoluble pesticides, regulates and slowdowns pesticide release, and increases bio-efficacy.

6.2 Use of ME in agriculture

Cyhalothrin ME was developed and found as the most promising pesticide formulation in China [23]. Carbendazim is another pesticide that was formulated as ME and its bio-efficacy was evaluated in *Rhizoctonia solani* [24]. Chlorpyriphos pesticide ME was successfully developed and found the most potent, safe, and environment-friendly pesticide formulations in comparison with conventional pesticide formulation in recent years [25].

Along with the synthetic pesticides, botanical-based microemulsions are well known in Asian agriculture. Essential oil in microemulsion form gives superior bioactivity as compared with emulsion forms [26]. In microemulsion system, spreading capacity and dispersion improves over applied plant surface. Therefore,

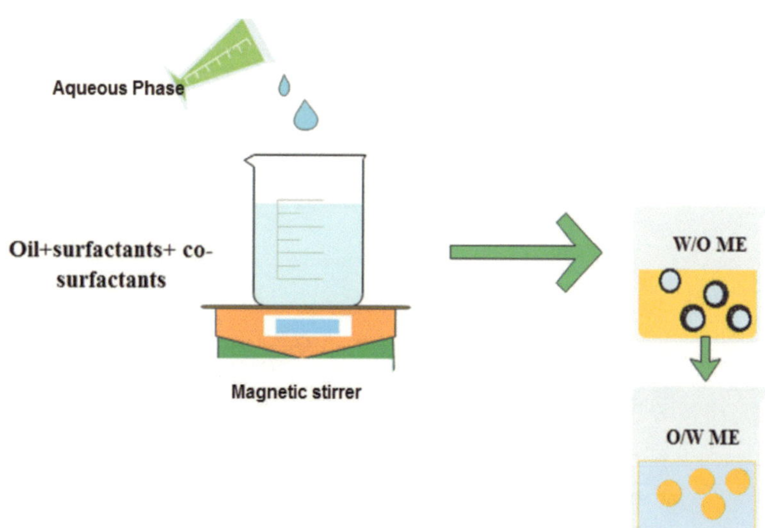

Figure 9.
Diagrammatic representation of the phase inversion method of microemulsion.

microemulsion system is the important vehicle for essential oil targeted delivery in a small quality without any losses [27, 28]. In addition, essential oil bio-constituents uniformly disperse after application over active targeted sites and offers improved bio-efficacy [29].

Clove (CO) and lemongrass oil (LGO) ME have investigated as efficient antifungal agents against *Fusarium oxysporum* f.sp. *lycopersici* without any phytotoxicity to main crop [30]. Previous studies in Asian region revealed that synthetic pesticide- and botanical-originated microemulsion could be an advanced, green, safe formulation against different crop pests. Therefore, microemulsion is the one of new-generation formulations for safest delivery system.

IPFT has developed different types of microemulsions of synthetic pesticides along with botanical pesticides. In addition to this, IPFT has prepared microemulsion system with inbuilt adjuvants and synergist, which will enhance the formulation efficiency and efficacy. There are different botanical-based microemulsions have been developed and formulation techniques transferred to various agrochemical industries.

6.3 Nanoemulsion formulation

Nanoemulsions are defined as nano-sized droplets dispersion in immiscible liquids (**Figure 10**). Different pesticides have been formulated as nano-formulation and researchers have quantitatively estimated the pesticide content by various characterization techniques [31]. This formulation exhibits the property of encapsulation and regulated release of pesticides for extended period of time as in controlled release formulations.

Characteristic features of nanoemulsion formulation techniques are small droplet size with low amount of surfactant and active ingredient. Similar to ME, NE can enhance solubility of active ingredient, increase bioactivity, and improve spreadability (during application). Moreover, it can reduce volatility and hydrolysis of active ingredient.

It has been evaluated that nanoemulsion formulation showed enhanced bio-efficacy results compared with emulsifiable concentrates (EC) and microemulsions (ME). This study was conducted in third-instar larva of *Plutella xylostella* [32]. Indian Agriculture Research Institute (IARI), India, has developed nano-sulfur

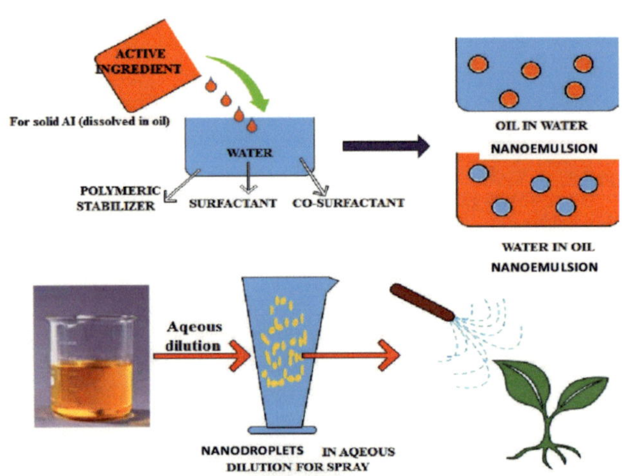

Figure 10.
Diagrammatic representation of nano-formulation preparation and its application.

formulation against *Erysiphe cichoracearum* (powdery mildew of okra) [33]. In addition to this, IARI has also developed nano-hexaconazole as effective fungicide [34]. Characteristic feature of this developed formulation is biosafety aspect, and after application of nano-hexaconazole, no impact was observed on soil nitrifiers such as blue green algae and cynobacteria species. Therefore, nano-formulations maintain the sustainable soil fertility and productivity compared with conventional formulations [35].

Many botanical-based nanoemulsions have been developed in recent years. In a study, *M. longifolia* oil nanoemulsion developed with droplet size 14 nm–36 nm. The results of study showed that in nanoemulsion formulation, its contact toxicity and durability increase. In a similar study, sea fennel (*Crithmum maritimum*) essential oil is formulated as nanoemulsion along with SiO2 nanoparticles and evaluated against *Spodoptera litura*. In other study, Eucalyptus oil NE prepared by emulsification method and insecticidal and repellent effect was evaluated against *Sitophilus oryzae*, *Rhizopertha dominica*, and *Tribolium castaneum* [36]. Nanoemulsion of *Piper aduncum* fruit extract has also been developed against cabbage pest *Crocidolomi apavonana*. Therefore, it has been concluded that botanical nanoemulsion formulation represents a new alternate for integrated pest management for organic farming promotion in Asian countries [37].

In addition to this, nanoemulsion formulation also been used as edible coating to improve the storability of fruits and vegetables in postharvest conditions for enhancing shelf life and prevention of microbial growth over fruits and vegetables [38].

IPFT has also contributed in developing many nanoemulsions of synthetic as well as bioactive pesticides and different essential oils for controlling various agriculture pests and micro-organisms. Besides this, IPFT has prepared nanoemulsion with botanical synergists and adjuvants to enhance the bioactivity and stability of nanoemulsion formulations. Moreover, combination of nanoemulsions is further developed for the amplified pest control applications in intense pest attacking conditions.

7. Emulsion in water formulation or oil in water emulsion (EW)

Emulsion in water (EW) formulation is suitable for liquid or a liquid or oily active ingredient. These formulations are dispersion of active ingredient in aqueous continuous phase (**Figure 11**). The size of the dispersed droplets ranges generally from 0.5 to 4–5 μm. EW formulations are obtained by high-shear emulsification process. Principally, EW formulations contain pesticides dispersed in the form fine liquid droplets in water and form oil-in-water (O/W) emulsions [39].

7.1 EW in agriculture

Lambda-cyhalothrin is broad-spectrum synthetic pesticide widely being used to control diamondback moth, cabbage caterpillar, cotton bollworm, and other pests that damage main food crops such as vegetables, soybeans, peanuts, and cotton [40]. This pesticide is commonly used in Asian countries due to its moderate toxicity, high insecticidal activity, and a long-lasting effect. However, lambda cyhalothrin previously available as EC and ME formulations pollutes the environment, costly due to loads of surfactants and consume non-renewable resources in the form of petroleum solvents [41]. Therefore, there is an urgent requirement to replace these shortcomings to further take the benefit of this broad-spectrum pesticide.

EW formulations such as limonene, peppermint oil, and spearmint oil have been developed and bio-efficacy evaluation was done on *Pseudococcus longispinus*.

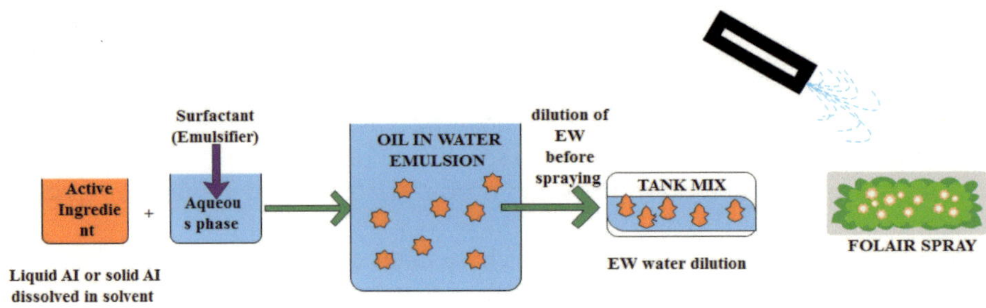

Figure 11.
Diagrammatical representation showing key steps of emulsion in water formulation and its application.

In a similar study, neem EW formulation was developed with palm oil methyl ester, RBD palm olein, and soybean oil for better adhesion and persistency. This formation was evaluated against golden apple snail. EW prepared with palm oil methyl ester showed better efficacies with LC_{50}, 45.30 mg/l under field condition and have longer persistence $t_{1/2}$ = 1.85; r^2 = 97.75 on paddy leaves [42]. The formulation was found to be very effective against the pest with 90% mortality [43]. In another study, bio-larvicide *Lagenidium giganteum* have been formulated as EW formulation. It was investigated that in EW formulation of *L. giganteum* mycelium shelf life and delivery improved for good and prolonged bio-efficacy. In a similar study, entomopathogenic fungi have been formulated as emulsion in water formulation and found to be very effective against various agricultural pests. The study concluded that EW formulation showed good bio-efficacy and stability of entomopathogenic fungi over unformulated form. Hence, emulsion in water is the safe and economical formulation against agricultural pest.

In addition to agricultural field pest control, EW formulation can also be used in preparation of postharvest packaging films with anti-insect property. Recently, cinnamon oil (CO) anti-insect packaging film has been developed for repelling *Plodia interpunctella* (Hübner) larvae [44].

IPFT has contributed to Lambda Cyhalothrin EW, Chlorpyriphos 10 EW, etc. Recently, IPFT has optimized the neem EW formulation procedure by high shear mixing. The increase in shearing intensity reduced the droplet size and resulted in higher stability.

7.2 Mixed formulations (Suspoemulsion, ZC, ZW)

The combination formulations have broad-spectrum insecticidal activities and can be applied for insect control in different Asian countries. These mixed formulations have the user and environment-friendly applications over conventional formulations.

1. ZW (Capsulated Suspension (CS) + Emulsion in water (EW)

This formulation is the combination of two formulations with two different pesticides in water medium. In this, one pesticide is encapsulated inside the polymeric coating and other is in emulsified droplet form (**Figure 12**).

IPFT has developed ZW combination formulation of capsulated suspension (CS) of Lambda cyhalothrin with concentrated emulsion in water (EW) of chlorpyriphos, and this combination was termed as ZW [7, 8]. Main advancement of this formulation is that it is the combination of two pesticides in two different formulations, one broad-spectrum pesticide, that is, in EW formulation for quick action and lambda cyhalothrin CS for controlled release and will be effective for extended

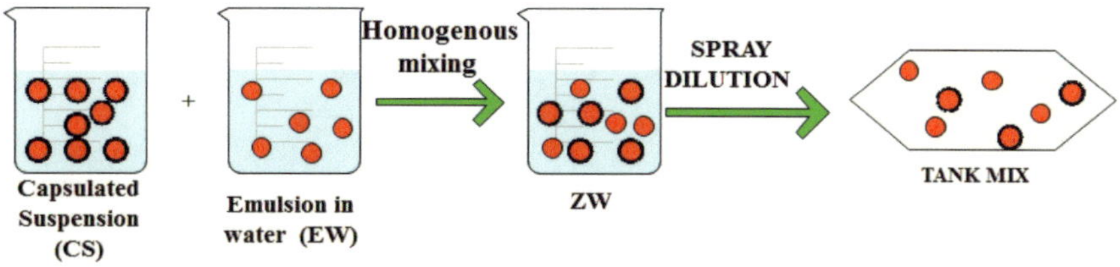

Figure 12.
Diagrammatical representation of ZC formulation and its application.

period of time and give long-term pest control [45]. This combination formulation can be used for different pesticides with good compatibility index.

2. ZC formulation (Capsulated formulation (CS) + Suspension Concentrates (SC))

The ZC formulation is the stable aqueous suspension of polymeric-encapsulated microcapsules and solid-suspended fine particles of two different pesticides. Both formulations are homogeneously mixed by wet milling and gentle shear mixing (**Figure 13**).

Chlorantraniliprole and thiamethoxam SC, lambda cyhalothrin and chlorantraniliprole ZC, thiamethoxam and lambda cyhalothrin ZC, beta-cyfluthrin and imidacloprid SC, and flubendiamide and thiacloprid SC efficacy have been evaluated against spotted pod borer. The study was found that ZC formulations gave superior results on the management of *M. vitrata* and *Spodoptera litura* over simple SC formulations [46].

In addition to enhanced bio-efficacy, this combination formulation has been investigated for non-targeted effects. ZC (thiamethoxam and lambda cyhalothrin under trade name Alika 247 bio-efficacy) was evaluated against Pest of Tea in West Bengal, India. The study reported that Alika 247 ZC was safe for the important natural predators found in the tea ecosystem relative to conventional formulations like EC [47]. Similarly, thiamethoxam and lambda cyhalothrin ZC impact was investigated on Population of Lady Bird Beetles in maize crop ecosystem in Gujarat, India [48].

This combination formulation has fast and quick knockdown and extended control of foliar insect pest. This formulation is basically developed for soybean aphids, Japanese beetle, grasshopper, corn rootworm beetle, stinkbugs, etc.

Institute of Pesticide Formulation Technology (IPFT) has developed Lambda cyhalothrin 14 CS with Diflubenzuron 10SC. The unique specialty of this developed ZC formulation is effective against early stages and adult stages simultaneously. The combined pesticide provides improved and synergistic activity. Besides this, formulation is suitable for immediate as well as for prolonged pest control practices.

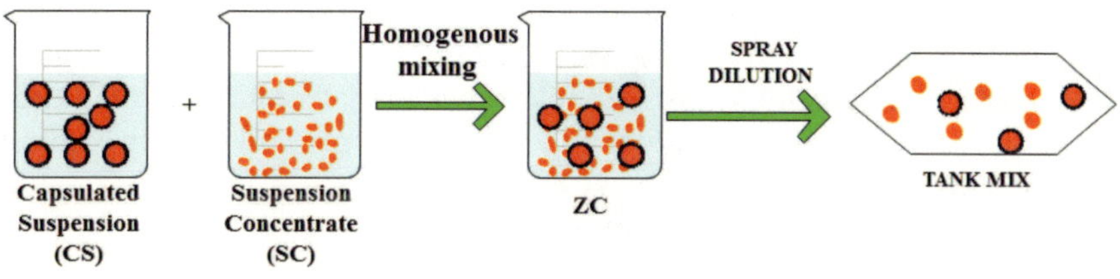

Figure 13.
Diagrammatical representation of ZC formulation and its application.

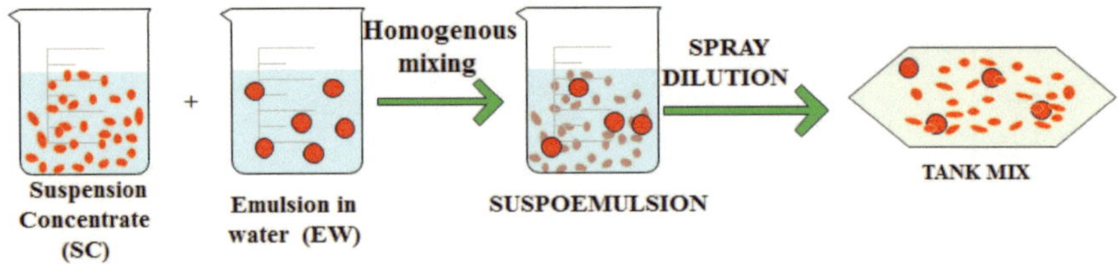

Figure 14.
Diagrammatical representation of suspoemulsion formulation and its application.

3. Suspoemulsion (Suspension concentrate (SC) + Emulsion in water (EW))

Suspension (SE) is the combination of two active ingredients one in suspension concentrate (SC) and concentrated aqueous emulsion (EW) (**Figure 14**). Suspoemulsion is the stable colloidal suspension with fine droplets with a high degree of electrostatic, steric, and hydrophobic interactions and with lesser degree of Ostwald ripening [49].

Main advancing features are different active ingredients with different solubility or melting points can be incorporated, providing broad-spectrum pest control, and tank mixing is not required. Suspoemulsions are the most convenient formulation for the farmers to apply the correct quantity of pesticides and tank mix incompatibility problems have been removed. Surfactants and thickeners were added in suspoemulsion to prevent flocculation and separation of the dispersed phases [50].

IPFT has contributed suspoemulsion of fipronil 5% SC + Soyabean oil as Adjuvant 5% EW. The main characteristic feature of this formulation is high stability and shelf life along with good bio-efficacy to agricultural pest in Asian countries.

7.3 Botanical formulations

Different bioactive phytochemicals have been identified for good bio-efficacy. These bioactive ingredients in formulated form will play a very significant role in promotion of organic farming in Asian countries in a safe and sound way. Followings botanical formulations have been formulated by IPFT.

New generation botanical-based formulations developed in IPFT, India, are as follows:

a. **Microemulsion:** *Neem oil-based microemulsions were successfully developed at IPFT, Gurgaon* [51]. *Different essential oil microemulsions have been developed by IPFT by using various botanical synergists and adjuvants.*

b. **Nanoemulsions:** Botanical origin nanoemulsions are very fine oil-in-water nano-droplet within the size of 5 nm–100 nm [52]. These nanoemulsions have both thermodynamic and kinetic stabilities [53]. IPFT has developed combined botanical nanoemulsion of eucalyptus oil with karanja and jatropha aqueous filtrates—for controlling stored grain pest *Tribolium castaneum* [54]. Uniqueness of this formulation is that karanja and jatropha aqueous extract was used from biodiesel waste product and with eucalyptus oil it gives combinatory activity. This formulation is the very efficient in terms of insect pest management and waste product management.

c. **Controlled release Formulations (CRF), microencapsulation***: This formulation technology regulates the release of pesticide and decrease the toxicity of pesticide* [55].

Highly volatile bioactive pesticides encapsulated in a thick polymeric coating by cross-linking. Therefore, these formulations are efficient for an extended period of time.

d. **Suspension concentrates:** Suspension concentrate (SC) botanical extracts have been developed by different researchers [56]. In botanical SC, botanical active ingredient is finely grinded and then dispersed in water medium with surfactants. Particle size distribution of diluted SC formulation is in the range of 2–20 μm. These formulations are eco-friendly and user friendly. Hence, botanical SC formulations are the most suitable formulation for botanicals to retain their greener characteristics.

e. **Oil dispersions (OD):** These formulations are similar to SC formulation only dispersing medium is oil in place of water as in SC. Therefore, OD formulations have good spreading and permeation compared with SC formulation. The oil dispersion formulation is the most suited formulation for hydrolytically unstable botanical pesticides. It has been reported that oil gives synergist action with botanicals and broadens the spectrum of pest management [57]. IPFT has developed oil dispersion formulations of many plant extracts against various insect pest. Recently, IPFT has developed tomato leaves extract OD formulation and found to be very efficient against mustard aphids.

Institute of Pesticide Formulation Technology (IPFT), Gurugram in India is the only institute devoted for the development of safe and environment-friendly new-generation insecticide formulation technology. There are some new formulated products of natural insecticides such as controlled release formulations, nano-formulations, or water-based formulations, which enhance the efficacy of natural pesticides against insect pest. The work carried out at IPFT greatly emphasizes on the development and promotion of environment and user-friendly pesticide formulations, also biodegradable, with the incorporation of latest technologies, and also on their commercialization.

8. Future considerations for the promotion of safe and green biopesticides

Currently, IPFT is working toward the development of safer alternatives to banned or going to be banned agrochemicals. Research is in the process of development to safer formulations with potentially low-risk user and environment-friendly novel formulation development of various broad-spectrum pesticides that are in the verge of banned have been attracting global attention. In this context, public and private sectors cooperation is necessary to facilitate the formulation development of safe and environment-friendly improved and advanced alternative. Novel formulations improve the delivery of agrochemicals and boost up the agricultural system in near future. Most important aspect along with the development is the cost of safe formulation. Maintaining low cost of novel formulations to farmers for a given product quality and availability, particularly in developing countries, is also important. Though, new formulation strategies could give out a very promising and potential option for pest control but to attain this objective, more field research is required to assess the efficacy on specific pest problems and over rising pest problems in various cropping systems. Therefore, it is a necessary requirement for strengthening the research in this safe and green formulation technology development.

Acknowledgements

Authors would like to thanks to Institute of Pesticide Formulation Technology under Ministry of Chemicals and Fertilizer, India. The authors would like to thank Dr. Saurabh Dubey for her valuable input in the preparation of this chapter.

Author details

Nusrat Iqbal[1]*, Amrish Agrawal[1], Md. Imteyaz Alam[2] and Jitendra Kumar[1]

1 Institute of Pesticide Formulation Technology, Gurugram, India

2 Department of Energy, Politecnico di Milano, Milan, Italy

*Address all correspondence to: nusratsiddiqa20@gmail.com

References

[1] Webster JPG, Bowles RG, Williams NT. Estimating the Economic Benefits of Alternative Pesticide Usage Scenarios: Wheat Production in the United Kingdom. Crop Production. 1999;**18**:83

[2] Schreinemachers P, Tipraqsa P. Agricultural pesticides and land use intensification in high, middle and low income countries. Food Policy. 2012;**37**:616-626

[3] Baskar K, Sudha V, Jayakumar M. Effect of Pesticides on Pollinators. MOJ Ecology & Environmental Science. 2017;**2**:40-52. DOI: 10.15406/mojes.2017.02.00052

[4] Damalas CA, Eleftherohorinos IG. Pesticide exposure, safety issues, and risk assessment indicators. International Journal of Environmental Research and Public Health. 2011;**8**:1402-1419. DOI: 10.3390/ijerph8051402

[5] Knowles A. Recent developments of safer formulations of agrochemicals. The Environmentalist. 2008;**28**:35-44. DOI: 10.1007/s10669-007-9045-4

[6] Meijs WHMA. New type of formulations and new application techniques; consequences for the authorisation of pesticides. Environmentalist. 2008;**28**:5-8

[7] Hazra D, Karmakar R, Poi R, Bhattacharya S, Mondal S. Recent advances in pesticide formulations for eco-friendly and sustainable vegetable pest management: A review. Archives of Agriculture and Environmental Science 2017a; 2: 232-237.

[8] Hazra DK, Karmakar R, Poi R, Bhattacharya S, Mondal S. Recent advances in pesticide formulations for eco-friendly and sustainable vegetable pest management: A review. Archives of Agriculture and Environmental Science. 2017b;**2**:232-237

[9] Gilden RC, HufflingK SB. Pesticides and Health Risks. Journal of Obstetric, Gynecologic, and Neonatal Nursing. 2010;**39**:103-110. DOI: 10.1111/j.1552-6909.2009.01092.x

[10] Hazra DK, Purkait A. Role of pesticide formulations for sustainable crop protection and environment management: A review. Journal of Pharmacognosy and Phytochemistry. 2019;**8**:686-693

[11] Li Z, Pang Y, Lou H, Qiu X. Influence of lignosulfonates on the properties of dimethomorph water-dispersible granules. BioResources. 2009;**4**:589-601

[12] Tian Q, Zhang Y, Jia Z, Zhang Q. Synthesis and hard water resistance mechanism of polycarboxylate dispersant for pesticide water dispersible granules. Journal of Dispersion Science and Technology. 2020;**41**(12):1892-1901

[13] Houbraken M, Senaeve D, Dávila EL, Habimana V, De Cauwer B, Spanoghe P. Formulation approaches to reduce post-application pesticide volatilisation from glass surfaces. Science of the Total Environment. 2018;**15**:728-737. DOI: 10.1016/j.scitotenv.2018.03.186

[14] Brown PW. Effects of particle size distribution on the kinetics of hydration of tricalcium silicate. Journal of the American Ceramic Society. 1989;**72**:1829-1832. DOI: 10.1111/j.1151-2916.1989.tb05986.x

[15] Zhang S, Yang X, Tu Z, Hua W, He P, Li H, et al. Influence of the hydrophilic moiety of polymeric surfactant on their surface activity and physical stability of pesticide suspension concentrate. Journal of Molecular Liquids. 2020;**317**:65-78

[16] Yan X, Zheng Z, Qi S, Rui Z, Guo-lin W, Xue-min W. Surfactant-enhanced phytoremediation of soils

contaminated with hydrophobic organic contaminants: Potential and assessment. Pedosphere. 2007;**17**:409-418

[17] Vineela V, Nataraj T, Reddy G, Vimala PS. Enhanced bioefficacy of Bacillus thuringiensis var. kurstaki against Spodopteralitura (Lepidoptera: Noctuidae) through particle size reduction and formulation as a suspension concentrate. Biocontrol Science and Technology. 2017;**27**(1):58-69. DOI: 10.1080/09583157.2016.1247433

[18] Peng R, Yuxia P, Xueqing Q, Yong Q, Mingsong Z. Synthesis of anti-photolysis lignin-based dispersant and its application in pesticide suspension concentrate. Royal Society of Chemistry Advances. 2020;**10**:13830-13837

[19] Devi PSV, Hari PP. Bacillus thuringiensis-based nanopesticides for crop protection. Journal of Biological Control. 2009;**23**:403-408

[20] Hao L, Lin G, Lian J, Chen L, Zhou H, Chen H, et al. Carboxymethyl cellulose capsulated zein as pesticide nano-delivery system for improving adhesion and anti-UV properties. Carbohydrate Polymers. 2020;**231**: 23-35

[21] Hazra D, Patanjali PK. Seed coating formulation technologies: An environmental biology friendly approaches for sustainable agriculture. Bioscience Methods. 2016. DOI: 10.5376/bm.2016.07.0005

[22] Sharma A, Dubey S, Iqbal N. Microemulsion Formulation of Botanical Oils as an Efficient Tool to Provide Sustainable Agricultural Pest Management [Online First]. Rijeka: IntechOpen; 2020. DOI: 10.5772/intechopen.91788

[23] (2009), Zhao, Feng, Hong-ying Xia, and Jing-ling He. "Formulation Design of cyhalothrin pesticide microemulsion." Current Science 97, no. 10 (2009): 1458-1462. http://www.jstor.org/stable/24107338.

[24] Leng P, Zhang Z, Li Q, Zhao M, Pan G. Microemulsion formulation of carbendazim and its in vitro antifungal activities evaluation. PLoS One. 2014;**10**:23-43. DOI: 10.1371/journal.pone.0109580

[25] Wu YQ, Xia HY, Zhao F. Research and exploitation of chorpyriphos pesticide microemulsion. Advanced Materials Research. 2013;**850**:1180-1183. DOI: 10.4028/www.scientific.net/amr.850-851.1180

[26] Balah MA. Formulation of prospective plant oils derived micro-emulsions for herbicidal activity. Mansoura Plant Protection and Pathalogy. 2013;**4**:911-926

[27] Giulia B, Marco C, Monica MF, Giovanni FP. Colloidal soft matter as drug delivery system. Journal of Pharmaceutical Sciences. 2009;**98**: 11-42. DOI: 10.1002/jps.21423

[28] Lawrence MJ, Rees GD. Microemulsion-based media as novel drug delivery systems. Advanced Drug Delivery Reviews. 2000;**45**:89-121. DOI: 10.1016/S0169-409X(00)00103-4

[29] Abeer H, Abdulaziz AA, Ramalingam R, Al-Bandari FA, Horiah AA, Dilfuza E, et al. Arbuscularmycorrhizal fungi regulate the oxidative system, hormones and ionic equilibrium to trigger salt stress tolerance in Cucumissativus L. Saudi Journal of Biological Sciences. 2018;**25**:1102-1114. DOI: 10.1016/j.sjbs.2018.03.009

[30] Sharma A, Sehra G, Sabnis P, Kamra A, Sharma S. Development of clove oil based nanoformulation against Fusariumoxysporum. Indian Phytopathalogy. 2016;**69**:313-315

[31] Kumar R et al. A simple method for estimation of sulphur in

nanoformulations by UV spectrophotometry. Current Science. 2011;**100**:1542-1546

[32] Feng J, Chen W, Shen Y, Chen Q, Yang J, Zhang M, et al. Fabrication of abamectin-loaded mesoporous silica nanoparticles by emulsion-solvent evaporation to improve photolysis stability and extend insecticidal activity. Nanotechnology. 2020

[33] Gogoi R, Singh PK, Kumar R, Nair KK, Alam I, et al. Suitability of nano-sulphur for biorational management of powdery mildew of okra (AbelmoschusesculentusMoench) caused by erysiphecichoracearum. Journal of Plant Pathology Microbiology. 2013;**4**:171-182. DOI: 10.4172/2157-7471.1000171

[34] Kumar R, Nair KK, Alam MI, Gogoi R, Singh PK, Srivastava C, et al. Development and quality control of nanohexaconazole as an effective fungicide and its biosafety studies on soil nitifiers. Journal of Nanoscience and Nanotechnology. 2015;**15**:1350-1356

[35] Kumar R, Gopal M, Pabbi S, Paul S, Alam MI, Yadav S, et al. Effect of nanohexaconazole on nitrogen fixing blue green algae and bacteria. Journal of Nanoscience and Nanotechnology. 2016;**16**:643-647. DOI: 10.1166/jnn.2016.10901

[36] Mohammed TGM, Nasr MEH. Preparation, Characterization and Biological Efficacy of Eucalyptus Oil Nanoemulsion against the Stored Grain Insects. Asian Journal of Advances in Agricultural Research. 2020;**13**:41-51. DOI: 10.9734/ajaar/2020/v13i230102

[37] Louni M, Shakarami J, Negahban M. Essential oil against Ephestiakuehniella (Lepidoptera: Pyralidae). Journal of Crop Protection. 2018;**7**:171-182

[38] Pirozzi A, Grosso D, Ferrari VG, Donsì F. Edible coatings containing oregano essential oil nanoemulsion for improving postharvest quality and shelf life of tomatoes. Foods (Basel, Switzerland). 2020;**9**:1605. DOI: 10.3390/foods9111605

[39] Liu Y, Wei FL, Wang YY, Zhu GN. Studies on the formation of bifenthrin oil-in-water nano-emulsions prepared with mixed surfactants. Colloids and Surfaces A: Physicochemical and Engineering Aspects. 2011;**389**:90-96

[40] Anjum F, Wright D. Relative toxicity of insecticides to the crucifer pests Plutellaxylostella and Myzuspersicae and their natural enemies. Crop Protection. 2016;**88**:131-136

[41] Sui H, Hua Z, Li X, Li H, Wu G. Influence of soil and hydrocarbon properties on the solvent extraction of high-concentration weathered petroleum from contaminated soils. Environmental Science and Pollution Research International. 2014;**21**:5774-5784

[42] Rosdiyani M, Ismail R, Siti N, Latip M, Hajjar SN, Latip. Efficacy and persistence of neem in emulsion-in-water (EW) formulation against the golden apple snail. 2016;**2**(9)

[43] Hollingsworth R, Hamnett RM. Using food-safe ingredients to optimize the efficacy of oil-in-water emulsions of essential oils for control of waxy insects. Acta Horticulturae. 2010;**880**:399-406. DOI: 10.17660/ActaHortic.2010.880.47

[44] Jo HJ, Park KM, Na JH, Min SC, Park KH, Chang PS, et al. Development of anti-insect food packaging film containing a polyvinyl alcohol and cinnamon oil emulsion at a pilot plant scale. Journal of Stored Products Research. 2015;**61**:114-118

[45] Takeshita T, Noritake K. Development and promotion of labor-saving application technology for paddy herbicides in Japan. Weed Biology and Management. 2001;**1**:61-70

[46] Reddy BKK, Paul A. Field efficacy of insecticide mixtures against the pod borer and leaf eating caterpillar in cowpea. Journal of Pharmacognosy and Phytochemistry. 2019;**8**:1224-1227

[47] Samanta B, Alam SKF, Patra S, Sarkar S, Dey PK. Alika 247 ZC (Thiamethoxam 12.6% + Lambda-Cyhalothrin 9.5%) against Pest Complex of Tea in West. Pesticide Research Journal. 2017;**29**:230-235

[48] Shailesh PD, Rajendra C. Impact of thiamethoxam 12.6% + lambda cyhalothrin 9.5% Zc on population of lady bird beetles in maize crop ecosystem. Advances in Life Sciences. 2016;**5**:153-158

[49] Memula S, Berger P, Chellappa C. Suspoemulsions with improved stability and correlation of long term stability with the zeta potential. in Pesticide Formulations and Application Systems: 15th Volume ed. H. Collins, F. Hall, and M. Hopkinson (West Conshohocken, PA: ASTM International, 1996), 132-144. 10.1520/STP16038S

[50] Tadros T, Cutler J, Pons R, Rossi P. Investigations of the Interaction Between Suspensions and Emulsions (Suspoemulsions). In: Ottewill RH, Rennie AR, editors. Modern Aspects of Colloidal Dispersions. Dordrecht: Springer; 1998. DOI: 10.1007/978-94-011-6582-2_22

[51] Singla M, Patanjali PK. Phase behaviour of neem oil based microemulsion formulations. Industrial Crops and Products. 2013;**44**:421-426

[52] Sadurní N, Solans C, Azemar N, García-Celma MJ. Studies on the formation of O/W nano-emulsions, by low-energy emulsification methods, suitable for pharmaceutical applications. European Journal of Pharmaceutical Sciences. 2005;**26**: 438-445

[53] Bouchernal K, Brianeon S, Perrier E. Nano-emulsion formulation using spontaneous emulsification, solvent oil and surfactant optimization. International Journal of Pharmacology. 2004;**280**:41-25

[54] Pant M, Dubey S, Patanjali PK, Naik SN, Sharma S. Insecticidal activity of eucalyptus oil nanoemulsion with karanja and jatropha aqueous filtrates. International Biodeterioration & Biodegradation. 2014;**91**:119-127

[55] Ribeiro C, Vicente AA, Teixeira JA, Miranda C. Optimization of edible coating composition to retard strawberry fruit senescence. Postharvest Biology and Technology. 2007;**44**:63-70

[56] Luckham PF. The physical stability of suspension concentrates with particular reference to pharmaceutical and pesticide formulations. Pesticide Science. 1989;**25**:25-34. DOI: 10.1002/ps.2780250105

[57] Vernner R, Bauer P. Q-TEO, a formulation concept that overcomes the incompability between water and oil. Pfalzenschutz-Nachrichten Bayer. 2007;**60**:7-26

Chapter 5

Characterization of *Rhizobium* and Plant Growth Promoting Rhizobacteria from French Bean Rhizosphere and Their Effect on French Bean Productivity

Saroj Kumar Yadav and Kiran P. Raverkar

Abstract

French bean (*Phaseolus vulgaris* L.) is used profusely by the common people as an alternative diet of protein. The sparse nodulation in French bean mainly may be due to lack of threshold level of specific rhizobial cells in soil at the time of sowing. The isolates streaked on YEMA with BTB changed to yellow color showing the production of acid which is the characteristic of *Rhizobium.* Utilization of different carbon sources is an efficient tool to characterize the isolates. Plant growth promoting rhizobacteria is the beneficial rhizobacteria inoculation of which increases growth and yield of French bean through different direct and indirect mechanisms. Inoculation of French beans with rhizobial and rhizobacterial isolates found to be improved growth, physiological, quality parameters and grain yield through symbiotic N_2-fixation capacity and plant growth promoting abilities. Co-inoculation of rhizobial and rhizobacterial isolates enhanced the growth and grain yield of French bean. These isolates may be used as consortium to improve the growth of French bean, which may reduce the dependency of farmer on chemical fertilizer as well as risk of pollution. In this chapter characterization of *Rhizobium* and plant growth promoting rhizobacteria and their effect on plant growth has been reviewed.

Keywords: *Rhizobium*, PGPR, biofertilizer, consortium

1. Introduction

In the present intensive agriculture practices leguminous plants play a critical role in natural ecosystem, agriculture and agroforestry because of their ability to fix nitrogen (N_2) in symbiotic relationship with *Rhizobium* and *Bradyrhizobium*. In addition to its role as a source of protein in the diet, biologically fixed N_2 is essentially free for use in economic terms by the host plant and by associated subsequent crops. This association improves the soil quality vis-à-vis sustainability. Among the legumes, French bean (*Phaseolus vulgaris* L.) is used profusely by the common people as an alternative diet of protein. It is very nutritious and contains 22.9 per cent protein, 1.2 per cent fat, 60.6 per cent carbohydrates and a large number of minerals like Ca (260 mg 100 gm^{-1} of seed) P (101 mg 100 g^{-1} of seed) and Fe

IntechOpen

(5.8 mg 100 g^{-1} of seed). French bean is sparsely nodulated throughout the India including North-West Himalayas putting it to disadvantage of biologically fixed-N_2 [1, 2] and thus responds to the enhanced levels of nitrogen [3]. In India, it is grown on an area of about 1 lakh hectare (ha) mainly in the states of Maharashtra (60,000 ha), Jammu and Kashmir (10,000 ha), Himachal Pradesh, Uttarakhand, Nilgiri (Tamil Nadu), Palni (Kerala) hills, Chickmagalur (Karnataka) and Darjeeling hills (West Bengal). The sparse nodulation in French bean mainly may be due to lack of threshold level of specific rhizobial cells in soil at the time of sowing. Recently different rhizobial strains have demonstrated various other plant growth promoting activities in addition to biological nitrogen fixation (BNF). This necessitates the isolation and development of the efficient multi-trait rhizobial isolates for French bean for economizing the nitrogen fertilizer, environmental safety and sustainable production.

Rhizobium plays a significant role in agricultural ecosystem services due to their ability to form symbiotic association with a wide range of leguminous plants that results in biological nitrogen fixation. Some of the rhizobial strains are reported to enhance the production of phytohormones, mineral uptake and reduce toxic effects of metals, thereby, indirectly promote growth and development of plant in polluted agricultural soils [4]. According to Tsai et al. [5] most commonly used French bean variety exhibit a high dependence on nitrogen fertilizers for growth and yield, and show considerable variation in their ability to nodulate and fix nitrogen, with the nitrogen percentage derived from atmosphere ranging from 68 to 72 per cent for the superior variety. This indicates that French bean needs starter nitrogen fertilization for sufficient nitrogen fixation. Inoculation of French bean with *Rhizobium* increases various plant growths, physiological, quality parameters and grain yield through symbiotic N_2-fixation capacity and plant growth promoting abilities. Plant growth promoting rhizobacteria is the beneficial rhizobacteria inoculation of which increases growth and yield of common bean through different direct and indirect mechanisms such as production of IAA, GA, HCN, Ammonia, siderophore and solubilization of phosphorus, potassium and zinc. Co-inoculation of *Rhizobium* with other PGPR; consortium enhanced the growth and grain yield of common bean [6]. By virtue of their rapid colonization of the rhizosphere and stimulation of plant growth, there is currently considerable interest in exploiting such microorganisms for enhanced crop yield. Therefore to harness the benefits of rhizobia and PGPRs in reducing the application of higher doses of inorganic fertilizers; the development of consortium comprising efficient multi-trait rhizobial isolates and efficient PGPRs are need of an hour for sustainable production of French bean. This chapter reviews the research reports relevant on the topic including (i) isolation and characterization of various isolates of French bean rhizobia (ii) authentication and evaluation of the efficacy of rhizobial isolates (iii) evaluation of the efficacy of potential multi-trait rhizobial isolates on growth and yield of French bean in relation to N dose (iv) isolation and characterization of PGPRs from the rhizosphere and the impact of various isolates of PGPR on growth of French bean (v) compatibility between selected rhizobial isolates and PGPRs; and development of consortium, and (vi) the interaction effect of efficient multi-trait *Rhizobium* and PGPR on French bean.

2. French bean and its use

Gramineae and Leguminosae are two major source of world's food supply, total 15 plant species of which account for more than 90 per cent of the total production of the major seed crops [7]. According to Harlan [8], three major cereal crops such as wheat (*Triticuma estivum* L.), maize (*Zea mays* L.) and rice (*Oryza sativa* L.) account for three quarters of the total food supply. The grains of these cereals provide carbohydrates for human and are complemented by the legumes [9] which vary in their carbohydrate and oil content but have high protein content [7]. In addition to important source of food, feed and fuel legumes are also a renewable source of nitrogen through atmospheric N_2-fixation for agriculture [10]. Incorporation of legume crops in field improves soil fertility and yield sustainability. Among the legumes, French bean (*Phaseolus vulgaris* L.), of American origin, is the most edible pulse in the world and is second only to the soybean (*Glycine max* L.) [11]. According to Broughton et al. [12], french bean (*Phaseolus vulgaris* L.) has been reported as an important legume for human nutrition and a major protein and calorie source in the world. It is cultivated in the sub-Himalayan and higher Himalayan altitudes between 1200 and 1800 m. In India, French bean covers an area of 2.3 mha with production of 1.1 million tonnes and productivity of 478 kg ha^{-1} [13]. French bean is popular among Indian farmers due to its high lucrative features such as short life cycle, good adaptability, high market value and the most important for poor farmers, particularly women, hence it is also known as woman's crop. French bean is a self-pollinating leguminous crop which belongs to the family Fabaceae and considered as an important crop in high population density areas of the world [14]. French bean is used both as a pulse and as a green vegetable [15]. In both the developed and the developing countries, French bean is consumed in different forms [16]. Seeds can be consumed as immature green grain. Dehulled seeds may be boiled, parched, roasted, germinated, fermented or cooked in different ways to suit specific tastes. In some parts of the tropics, the young leaves are used like spinach. Common bean seeds are also cooked with tomato sauce and canned. The residual straw can be used as fodder and forage [16] as well as to incorporate in the soils to improve the soil health.

3. Adaptation

French beans are well adapted to tropics, subtropics, and warm temperate regions, grown from 40°S to 40°N latitude. French bean completes their life cycle within 80 to 110 days which depends on variety and night temperatures during the growing season. Suitable temperature for growth of French bean varies between 20 and 22°C. The maximum temperature during flowering of French bean must be under 28°C. It requires a minimum of 500 to 600 mm of rain during the growing season if the crop is cultivated under rainfed conditions whereas an annual total of 600 to 700 mm is considered ideal. They are planted in warm soils with minimum temperatures preferably above 15°C after all danger of frost has passed. Soil texture such as sandy loam, sandy clay loam or clay loam with good drainage and clay content between 15 and 35 per cent is supposed to be best for cultivation of this crop. Soil pH of 6.0 to 6.5 is considered to be the best for the cultivation of French bean.

4. Biological nitrogen fixation

Biological N_2-fixation is a biological phenomenon, which involves some legumes, whether grown as pulses for seed or as pasture in agro-forestry or in natural ecosystems [17]. Biological nitrogen fixation is very efficient in satisfying the high nitrogen requirements of legumes because of the conversion of gaseous nitrogen (N_2) to ammonia (NH_3) making it available to plant use. Enzyme nitrogenase facilitated the process of BNF. Many N_2-fixing prokaryotes are diazotrophic, *i.e.* they can grow using dinitrogen gas as their sole source of N while other organisms can fix N_2 only in symbiosis with another eukaryotic organism. The equation for the reaction is.

$$N \equiv N + 8H^{+} + 8e^{-} + 16ATP \rightarrow 2NH_3 + H_2 + 16\,ADP + 16\,Pi \qquad (1)$$

Two protons are reduced by hydrogen for fixation of one molecule of dinitrogen and because of high stability of dinitrogen the reaction needs high energy [18]. Dupont et al. [19] reported that soon after germination of legume seeds, rhizobia present in the soil or added as seed inoculum invade the root hairs and move through an infection thread to the root. The bacteria multiply rapidly in the root, causing the swelling of root cells to form nodules. Nitrogen in the air of soil pores around the nodules is fixed by binding it to other elements and thus changing it into a plant available form. Some of the carbohydrates manufactured by the plant photosynthesis process are transported to the nodules where they are used as a source of energy by the rhizobia. The rhizobia also use some of the carbohydrates as a source of hydrogen in the conversion of atmospheric nitrogen to ammonia. Despite BNF being a naturally occurring process, many soils do not harbor sufficient numbers of appropriate rhizobia for effective symbioses. Inoculation of leguminous crop with appropriate and compatible rhizobia ensures maximum BNF. Inoculation is generally needed when certain new leguminous crops are introduced to new areas.

5. *Rhizobium*

Rhizobiaceae family is a physiologically heterogeneous and genetically diverse group of soil organisms, which are called rhizobia [20]. Rhizobia include a group of soil bacterial genera viz.; *Rhizobium*, *Bradyrhizobium*, *Sinorhizobium*, *Mesorhizobium*, *Allorhizobium* and *Azorhizobium* which have ability to nodulate symbiotically the members of the plant under Leguminosae family [21, 22]. *Rhizobium*-legume associations are very specific therefore the nodules will be formed in the legume only when infected with a specific *Rhizobium* [23]. According to Broughton et al. [12] specificity involves the recognition of the bacterium by the host and of the host by the bacterium through the exchange of signal compounds which induce differential gene expression in both partners. The bacteria which are able to form root nodules in French bean have been classified into five species of the genus *Rhizobium*, *R. leguminosarum* biovar (bv.) *phaseoli* [24], *R. tropici* [25], *R. etli* bv. *Phaseoli*, *R. gallicum* bvs.*Gallicum* and *phaseoli* and *R. giardinii* bvs.*Giardinii* and *phaseoli*. From various studies, it has been observed that *Phaseolus* rhizobia are very diverse at the species, intra-species and population levels. According to Aguilar et al. [26], current evidence refers difficulty to recognize factors which involved in the distribution of the different rhizobial species among sites, although there is increasing evidence in the literature of parallel evolution between bacteria and the French bean.

6. Nitrogen fixation by *Rhizobium*

French beans have low ability to fix nitrogen symbiotically and surprisingly larger rates of N_2-fixation can be obtained under appropriate conditions [27]. In the light of the poor nodulation in French bean, in general in India, it is feasible that under these situations BNF technologies can become extremely important in order to reduce the use of chemical nitrogenous fertilizers, improve the soil health and enhance the yield levels. Hence, inoculation with the effective rhizobial inoculum presents a great potential for increasing food production in N-W Himalayas and other parts of the French bean growing area. The number of nodules in the plant decreases with the higher rates of soil N application at planting. In leguminous plants, Nitrogen fixation is a symbiotic process between nitrogen fixing bacteria and legume roots, and occurs within specialized root nodules. Hungria et al. [28] observed an adverse effect on leguminous root nodule development at low temperature stress.

7. Morphological and biochemical characteristics of rhizobia

Morphological characteristics of rhizobia refers to external appearance of rhizobia *viz*; shape, size and color; and biochemical characteristics refers to different characteristics which are produced by rhizobia through their chemical and microbial activities such as acid or alkali production, CRYEMA test, GPA test, carbohydrate utilization, enzymatic activity and plant growth promoting traits. Yadav et al. [29] studied 50 rhizobial isolates separated from the nodules of French bean (*Phaseolus vulgaris* L.) were tested and exhibited typical characteristics of *Rhizobium* sp. on yeast extract mannitol agar media supplemented with Congo red. In Kenya, genetic characterization and diversity of *Rhizobium* isolated from root nodules of climbing bean (*Phaseolus vulgaris* L.) varieties were studied by Koskey et al. [30] and they found that none isolates absorb Congo red dye when incubated in the dark on CRYEMA medium found Gram +ve rods. All isolates were found to be acid producers and fast growers by turning BTB indicator from deep green to yellow when grown on YEMABTB. Most of the isolates showed a mucoid texture because of the exopolysaccharides production. In the study of biochemical characterization of French bean associated rhizobia, Rai and Sen [31] observed and reported that colonies of *Rhizobium* were circular, convex, semi-translucent, raised, single and mucilaginous in nature. According to Vincent [27] and Holt et al. [32] the colonies were large (2–4 mm in diameter) mucilaginous, circular, convex with smooth edges, glistening translucent or white and precipitated calcium glycerophosphate present in YEM agar. *Rhizobium* test in Congo red showed that the colonies did not absorb the congo red color which differentiates *Rhizobium* from *Agrobacterium* [33]. According to Deka and Azad [34] *Rhizobium* cannot grow in Hoffer's medium, however; in contrast Rai and Sen [31] studied and observed that few of the isolates like S-3, CBR and K-1 showed mild growth. The growth of *Rhizobium* in the Hoffer's medium was also observed by Dubey et al. [1]. Deshwal and Chaubey [33] observed no yellow zone around the colonies of *Rhizobium* and such negative ketolactase activity confirmed the isolates to be free from any contamination of *Agrobacterium*.

The isolates changed to yellow color showed the production of acid which is the characteristic of *Rhizobium* [35]. Similarly, isolates of *Rhizobium leguminosarum* bv *trifolii* associated with clover showed growth and turned the yeast extract mannitol agar media containing BTB to yellow color indicated all were fast growers and acid producers. It was reported that the utilization of glucose as a carbon source is a confirmatory test for *Rhizobium* [35]. Utilization of different carbon sources is an efficient tool to characterize the isolates [36]. Only four isolates obtained in the study were able to use dextrin as a carbon source, which is in accordance with other works indicating that dextrin is rarely utilized by *Rhizobium* [24, 31]. The utilization of majority of carbon and sodium organic salt sources by *Rhizobium* has also been reported [37].

In the glucose-peptone agar medium, growth of the *Rhizobium* has been observed by [1]. Hunter et al. [38] observed the negative gelatinase activity which is a feature of *Rhizobium*. Yellow slants and red butt were obtained showing the utilization of glucose and sucrose in the triple sugar iron agar medium [35]. De Oliveira et al. [39] also observed that *Rhizobium* strains obtained from different sources can utilize starch. Rhizobial isolates may not grow on lactose [35]. As the pH becomes high, color of the media changes from yellow to pink which indicates the production of ammonia because of urease enzyme secretion by the incubated isolates which is a positive reaction for the test [40]. Gauri et al. [37] observed that all isolates of rhizobia showed a positive test for urease.

Biochemical characterization and protein profile by sds-page of French bean (*Phaseolus vulgaris* L.) associated rhizobia conducted by Kumari et al. [41] in Andhra Pradesh. They isolated total of six isolates. All the rhizobial isolates were positive to the indole acetic acid, nitrate reduction, urease, catalase and oxidase. Some of the isolates did not produced H_2S and consumed citrate as a sole source of carbon. Positive results were found from the starch hydrolysis assay. On subjecting inoculated plates to iodine test, clear zones from place to place were observed and the colonies changed to yellow color, however blue color appeared on no growth areas. It designates the isolates have the potential to hydrolyze starch present in the medium.

8. Morphological and biochemical characteristics of PGPR

The plant growth promoting rhizobacteria must be defined by some important attributes such as (a) an efficient tool to colonize the root surface (b) to survive, multiply and compete with other microorganisms and (c) to promote plant growth [42]. Anitha and Kumudini [43] reported that the isolates of fluorescent *Pseudomonas* on King's B agar produced creamy, convex colonies having 1–2 mm diameter and at 265 ηm appeared yellowish-green fluorescence. Microscopic studies showed that the isolates were gram negative and rod shaped. These isolates were found to utilize sucrose, mannitol and lactose to varying extent and showed that the bacterial isolates were positive only for catalase, oxidase, organic acids, citrate, amylase, indole and caseinase. According to Battu and Reddy [44] gram negative and rod shaped colonies, produced yellowish green pigment on King's B medium were positive for gelatinase and oxidase which were identified as *Pseudomonas flurescence*. Rodríguez-Cáceres [45] determined morphology and motility for each isolate and also performed biochemical tests such as nitrate reductase and urea hydrolysis [46]. Qualitative analysis showed that all bacterial isolates produced IAA, ammonia, siderophore and hydrogen cyanide.

9. Authentication and evaluation of *Rhizobium*

Authentication is the process by which we can ascertain that the isolates are *Rhizobium* or not through its infection capability under gonotobiotic conditions. Authentication of rhizobia to determine their symbiotic efficiency is required to screen out effective native rhizobial isolates [47]. In order to achieve maximum legume productivity, screening of native isolates for their N_2-fixation efficiencies [48] is important for the development of effective legume inoculum. Nodulation ability and effectiveness of native rhizobia from the seven districts of Uttarakhand in French bean was determined by Yadav et al. [29] and reported that out of fifty isolates 36 were authenticated as *Rhizobium* based on their ability to nodulate French bean. Morphological assessment and effectiveness of indigenous rhizobial isolates nodulating *Phaseolus vulgaris* in water hyacinth compost testing field in Lake Victoria basin was studied by Muthini et al. [49] and reported that the isolates obtained in his study had the ability to renodulate (Infectiveness) *Phaseolus vulgaris* under bacteriologically controlled conditions. Bala et al. [50], who reported that appropriate rhizobial isolates nodulate and fix N_2 on the target host and that each isolate was able to form nodules with the host plant was identified as *Rhizobium*. High degree of symbiotic efficiency of the specific indigenous strain $S_{21/6}$ was recorded indicating different symbiotic potential of indigenous strains and confirmed the importance of rhizobial strain selection [51].

Evaluation or screening of the authenticated isolates can be done on the basis of plant dry matter response (effectiveness), nodule number and nodule dry weight of the inoculated isolates. Isolation, authentication and evaluation of rhizobial isolates from the soils of North-West Himalayas in French Bean (*Phaseolus vulgaris* L.) studied by Yadav et al. [29] and reported that French bean inoculation with the rhizobial isolates significantly increased the dry shoot, root and total biomass; and shoot: root ratio. The inoculation of French bean with rhizobial isolate RA6 produced total dry biomass significantly higher by 154.4 per cent over reference strain i.e. MTCC 10096. This results demonstrated the presence of native rhizobia in soils of N-W Himalayas capable of nodulating French bean that are either superior or at par with reference strain in improving the overall growth and synthesis of higher biomass in French bean. Muthini et al. [49] reported that nodulated plants had higher shoot dry weight, than the non nodulated plants, however, the mean shoot dry weights was not directly related to the nodule number or nodule dry weight.

10. *Rhizobium* inoculation, N-levels and interaction impact

10.1 Rhizobium

Grain legumes have been recognized worldwide as an alternative means of improving soil fertility through their ability to fix atmospheric nitrogen, increasing soil organic matter and improving soil structure [52]. Several studies have sought to identify efficient and competitive strains of rhizobia to cope the nitrogen requirements of common bean [53]. Deshwal et al. [54] observed that in addition to BNF rhizobia can promote plant growth by different direct or indirect mechanisms such as production of IAA, GA, solubilization of inorganic phosphates and biocontrol of plant diseases. Beneficial effects of rhizobium on French bean have been reported by various workers under different climatic and soil conditions [55, 56]. Ndlovu [57] reported that nodulation was significantly affected by inoculation with *Rhizobium phaseoli*. Nodule dry biomass plant^{-1} was significantly increased by approximately 51.11 per cent with inoculation compared to uninoculated treatment in 2012–2013. Several workers reported that inoculation of legumes with *Rhizobium* isolates improved nodulation and had a positive effect on a number of plant growth parameters [58]. Das [59] also reported that higher number of nodules plant^{-1} was observed in inoculated plants than the un-inoculated. This might be due to application or introduction of inoculants that increased number of the *Rhizobium* bacteria which infect the roots to form nodules. The higher number of bacteria resulted in higher number of vigorous nodules plant^{-1}. The number and size of nodules indicated the amount of plant tissue available for nitrogen fixation. Thus, the results of this study also suggested a good symbiotic association between *Rhizobium phaseoli* and the host French bean. The presence of nodules in uninoculated treatments during both seasons might be due to the result of existing indigenous *Rhizobium* present in the soil. The increase in dry biomass of nodules which was formed by the inoculation with *Rhizobium phaseoli* might be results of efficiency of the strain.

There were significant differences in shoot, root and total dry biomass of *P. vulgaris* inoculated with variable rhizobial isolates. The significant differences in the shoot dry biomass showed clear differences in the ability of the isolates to fix nitrogen and are among the preferred methods for determining symbiotic effectiveness of rhizobial isolates [3]. According to Meena et al. [60], application of *Rhizobium* significantly increased the plant height, germination, number of branches plant^{-1}, number of leaves, leaf length and leaf width. The improvement in plant height and dry biomass production at flowering as well as at harvest might be due to the plant growth promoting capabilities, carbohydrates utilization abilities and improved nodulation. The differences in plant growth due to rhizobial inoculation have been attributed to changes in assimilate partitioning [58]. The higher dry shoot biomass of common bean by *Rhizobium* inoculation seems to be due to the supply of N to the crop through symbiotic N_2-fixation [61].

Yadegari and Rahmani [62] recorded the positive effect of inoculation with rhizobial strains Rb-133 and Rb-136 on plant growth. Rhizobial strain increased the seed yield, number of pods plant^{-1}, number of seeds pod^{-1}, weight of 100 seeds, seed protein yield, total dry matter over uninoculated control plants. During two years of study they registered the seed yield in inoculated plants ranging from 1221 to 4693 kg ha^{-1} depending on the strain and cultivars. Inoculation with suitable strains of *Rhizobium* has been recognized as prerequisite for increasing the yield and quality of legumes [63].

Inoculation of seeds by *Rhizobium* sp. prior to planting has also been reported to be a key factor in enhancing nodulation, early emergence, crop vigor and high grain yield [64, 65]. Bambara and Ndakidemi [66] also reported high common been seed yield of 1679 kg ha^{-1} with inoculation compared to 758 kg ha^{-1} from the uninoculated control. The inoculation of seeds with *Rhizobium* increased nodulation, protein and chlorophyll content, nitrogen uptake, growth and yield parameters of legume crops [67]. In Iran, Namvar et al. [67] reported that *Rhizobium* inoculated plants showed more chlorophyll content and LAI than uninoculated plants. *Rhizobium* inoculation increased chlorophyll content and LAI by 5.43 and 6.99 per cent, respectively as compared to uninoculated plants. In West Bengal, varietal performance of bush type French bean (*Phaseolus vulgaris* L.) for growth, fresh pod yield and quality was studied by Das [59]. Their results revealed that *Rhizobium* inoculation increased the yield and quality parameters *viz*; protein content, vitamin-A content and ascorbic acid content in the fresh pods of the French bean varieties. Under *Rhizobium* inoculation special jhhati beans recorded higher pod yield (23.05 t ha^{-1}) over uninoculated control (20.05 t ha^{-1}). Meena et al. [68] observed and reported, among the six different biofertilizers the best biofertilizer B3 (*Rhizobium*) is recorded significantly improvement in various yield and quality traits. Higher yields obtained with inoculation confirm that the *Rhizobium* technology is efficient in supplying nitrogen to legumes and is a better option for resource-poor farmers who cannot afford to purchase expensive inputs as well as potential strategy to nullify the adverse impact of chemical fertilizer on environment.

Number of pods, number of grains, grain yield, protein content and protein yield in French bean was influenced significantly due to inoculation with rhizobial isolates over uninoculated control [69]. This positive effect due to inoculation with rhizobial isolates attributed to nodulation and nitrogen fixing capability which enhances the nitrogen supplement to plants, resulted higher vegetative growth and carbohydrate portioning and more protein formation. Phosphorus solubilizing capability of rhizobial isolates helps to solubilize inorganic fixed phosphate and make it available for plant uptake. Higher availability of phosphorus improves N_2- fixation, root proliferation which results higher uptake of nutrient from soil and phosphorus also acts synergistically with nitrogen and helps in carbohydrate translocation and protein synthesis. An increase in number of pods plant^{-1}, number of grains pod^{-1} and pod yield due to *Rhizobium* inoculation might be due to more availability of nitrogen inside the plant bodies [59]. Koskey et al. [30] reported that inoculation of climbing beans in the field and green house significantly enhanced nodule and shoot dry biomass, number of pods plant^{-1}, seed yields and nitrogen content in shoot of MAC 13 and MAC 64 climbing beans. Ali (1998) also reported the same results and he concluded that inoculation increased pod yield plant^{-1}. With respect to *Rhizobium* inoculation treatment, higher pod yield was recorded in *Rhizobium* inoculation (20.73 t ha^{-1}) compared to without *Rhizobium* inoculated plants (17.68 t ha^{-1}). Increase in total pod yield is due to more nitrogen availability in the inoculated plants. The higher values of protein content are estimated in *Rhizobium* inoculated plants (2.09%) in comparison to without *Rhizobium* inoculation (1.44%). Increase in protein content due to *Rhizobium* inoculation might be the result of increased nitrogen content inside the plant body which is main element for protein synthesis.

10.2 Nitrogen levels

Nitrogen is an essential nutrient for plant growth and development. Nitrogen deficiency is frequently a major limiting factor for crop production all over the world [70, 71]. Therefore, adequate supply of nitrogen is necessary to achieve high yield potential in plants which usually depend upon combined or fixed form of N such as NH_4^+ and NO_3^- because it is unavailable in its most prevalent form as atmospheric N. The sparse nodulation in French bean needs more amount of nitrogen for growth and development in comparison to other legumes.

In French bean all the plant growth parameters except nodule number and nodule biomass were significantly improved with higher level of nitrogen application (from 0 to 120 kg N ha^{-1}), reported in several studies (**Table 1**). This improvement is attributed to the high vegetative growth and higher formation of photosynthates. Nitrogen plays an important role in the formation of protein and nucleic acids structure, the most important building material for every cell. In addition to it nitrogen is also a component of chlorophyll that enables the plant to capture energy from sunlight thorough photosynthesis. Thus, nitrogen supply to a plant increased the concentration of protein, amino acids, protoplasm and chlorophyll which influenced cell size, leaf area and photosynthetic activity [91, 92]. Increased level of nitrogen application in French bean resulted in increased plant height [81]. The negative effect of N fertilizer on French bean nodulation is well documented [23]. However, farmers have gradually adopted the use of N fertilizers with French bean crops, to maximize yields, particularly when irrigation is used. According to Yadav [69] number of trifoliate leaves, leaf area and chlorophyll content of French bean was significantly increased with higher levels of nitrogen (from 0 to 120 kg N ha^{-1}), the reason behind that N is chief constituent of amino acids which is the building unit of protein, protoplasm leading to improved vegetative growth. Nitrogen being constituent of chlorophyll higher dose of nitrogen increases chlorophyll concentration resulted in more photosynthesis and enhanced number of trifoliate leaves and leaf area which was reported by various workers (**Table 1**). N fertilization upto 120 kg N ha^{-1} in French bean increased number of pods plant^{-1} [77–79, 85]. Nitrogen supply affects a wide range of physiological processes in higher plants [87].

10.3 *Rhizobium* x N-level interaction

An inoculation of rhizobial isolates in combinations with different levels of nitrogen significantly improved the various plant growth parameters in French bean as compared to uninoculated control [61, 69, 90, 93, 94]. Growth, symbiotic and yield response of N-fertilized and *Rhizobium* inoculated common bean (*Phaseolus vulgaris* L.) was conducted by Yoseph and Shanko [94] at Hawassa University, Ethiopia and reported that N fertilization and *Rhizobium* inoculation had significant effect on

Growth parameters	References
Plant height	Negi and Shekhar [72]; Dwivedi et al. [73]; Kushwaha [74]; Saxena and Verma [75]; Singh and Rajput [76]; Singh and Singh [77]; Dhanjal et al. [78]; Prajapati et al. [79]; Veeresh [80], Jagdale et al. [81]
Number of branches plant^{-1}	Dhanjal et al. [78]; Prajapati et al. [79]; Veeresh [80]
Dry biomass production	Veeresh [80]; Prajapati et al. [79]; Singh et al. [82]
Number of pods plant^{-1}	Sharma et al. [83]; Rajput et al. [84]; Singh and Singh [77]; Behura et al. [85]
Number of seeds pod^{-1}	Dhanjal et al. [78]; Prajapati et al. [79]; Veeresh [80]; Behura et al. [85]; Singh et al. [82]
Grain yield	Singh and Singh [77]; Singh et al. (2000); Dhanjal et al. [78];Veeresh [80]; Singh et al. [82]
Physiological parameters	
Number of trifoliate leaves	Escalante et al. [86]; Cechin and Fumis [87]; Yadav [69]
Leaf area plant^{-1}	Hegde and Srinivas [88]; Rana and Singh [89]; Escalante et al. [86]; Veeresh [80]; Namvar et al. [90]
Chlorophyll content	Escalante et al. [86]; Rana and Singh [89]; Namvar et al. [90]
Quality parameters	
Protein content	Verma and Saxena [75]; Singh and Singh [77]; Veeresh [80]
Water soluble carbohydrate	Namvar et al. [90]; Yadav [69]

Table 1.
List of plant growth, physiological and quality parameters of French bean reported by various workers.

plant height, dry shoot weight, nodule number plant^{-1}, nodule dry weight plant^{-1}, number of pods plant$^{-1,}$ number of seeds pod^{-1} and grain yield. Omoregie and Okpefa [95] observed that when initial levels of available soil nitrogen were low, a period of nitrogen hunger can reduce nodulation. Kucuk [96] studied the effect of *Rhizobium* inoculation either alone or in combination with nitrogen applications on French bean (*Phaseolus vulgaris* L.) and reported that plant heights were significantly affected by the control, nitrogen, inoculation and different treatment x variety interactions. Interaction effect between nitrogen and rhizobial isolates significantly improved plant height, number of trifoliate leaves, dry biomass production, and leaf area, chlorophyll content, nitrogen status, grain yield, protein content and protein yield. This could be attributed to additional external supply of nitrogen for vegetative growths of plants and to be used by microbes to fulfill its requirement as starter in initiation of nitrogen fixation. Combined application of rhizobia and nitrogen may be beneficial for improvement of root proliferation and plant growth due to plant growth promoting ability of rhizobial isolates which need more nitrogen for their metabolic activity, might be fulfilled through external supply. Generally, inoculation with *Rhizobium* at all levels of nitrogen application increased biomass production over uninoculated plants [61, 67, 97]. Sajid et al. [98] concluded that the *Rhizobium* inoculation produced higher grain yield than without inoculation. It might also be due to more number of pods and seeds due to *Rhizobium* inoculation and applied N.

11. Compatibility between *Rhizobium* and PGPR

Under natural soil conditions microorganisms are effective to colonize the plant roots for function. Compatibility between the PGPR and other microorganism to colonize the root system without inhibiting each other is a prerequisite for getting beneficial result using multiple microbes in a crop field. The use of mixed cultures of beneficial microorganisms as soil inoculants is based on the principles of natural ecosystems which are sustained by their constituents such as the quality and quantity of their inhabitants and specific ecological parameters [99]. In the rhizosphere, PGPR and nodule promoting rhizobacteria induce phytoalexins production by the plant, creating antibiosis in the rhizosphere for pathogenic forms, siderophores production to chelate insoluble cations and associative action with the plant [100, 101]. *Rhizobium* as a gram -ve bacteria, is able to establish symbiosis with leguminous plants such as *Cicer* as well as many other rhizobacterial strains, and develops positive interactions with legumes by inhabiting root nodules. Within these nodules, bacteria reduce atmospheric nitrogen to ammonia which acts as a sufficient useable nitrogen source [102]. Studies on legume rhizosphere bacteria have shown that besides indigenous rhizobia interacting and competing for nodulation with an inoculant strain by antagonistic or synergistic interactions, other diazotrophs such as *Azotobacter* and *Azospirillum* as well as rhizosphere fungi and bacteria especially species of *Pseudomonas* and *Bacillus* do interact with *Rhizobium* affecting nodulation and nitrogen fixation [58].

12. Impact of PGPR on French bean

The effect of Plant growth promoting rhizobacteria on plant growth is a well-documented fact. PGPR plays an important role in agricultural systems, especially as biofertilizer. A positive influence of inoculation with various PGPR isolates on shoots and roots length; dry biomass production; and shoot: root ratio was studied by Yadav [69]. The higher shoot length, root length, root volume and dry biomass production due to inoculation with various PGPR isolates could be attributed to their plant growth promoting traits such as IAA, GA, P-Solubilization, Zn-Solubilization, ammonia production, HCN production and nitrogen fixation. IAA and GA are the plant growth hormones in which IAA controls processes *viz*; differentiation, division and enlargement of cells which controls plant growth whereas GA plays pivotal role in growth and development of plants. In most of the observed studies, the growth controller especially IAA, influences most of the root system like primary root growth, side root and piliferous layer formation [103] helps in plant growth promotion. The significant increase in growth of shoot and root due to inoculation of isolates indicates that the bacterial isolates have ability to provide better nutrient flux to the plant host which resulted in the increase of the plant biomass and N accumulation. Beneficial responses of PGPR due to beneficial interaction with rhizobia on legumes have been reported by various workers [75]. An increase in shoot: root ratio of French bean due to inoculation with the PGPRs isolates indicated that carbohydrate might be translocated to shoot but increase in root: shoot ratio in plant due to inoculation with isolates indicated more accumulation of carbohydrates in root rather than its portioning. High root: shoot ratio of plant indicated that plant may be survived in water or salinity stress conditions. The high root: shoot ratio due to inoculation of PGPR was also reported by [69]. Kloepper [42] reported that various PGPR isolates can alter the root architecture and promote plant development through the synthesis of different phytohormones such as IAA, GA and cytokinin. Stefan et al. [104] studied the effects of inoculation with PGPR on photosynthesis, antioxidant status and yield of runner bean and reported that PGPR strains used for seed inoculation induced significant increase in photosynthetic rate at 20 DAI. Increased photosynthetic activity is a consequence of a higher N incorporation which contributed to the formation of chlorophyll [105].

Inoculation of PGPR strains increased the nutritive value of grains by increasing the soluble protein content up to 16.24 per cent and total reducing carbohydrates concentration up to 49.28 per cent.

13. Interaction effect of *Rhizobium* and PGPR on growth of French bean

An application of PGPR together with *Rhizobium* improved the growth and seed production of beans [62]. Plant growth parameters, physiological parameters and quality parameters of French bean were influenced due to consortium comprising of efficient rhizobial and PGPR isolates [69]. The plant growth parameters were significantly improved due to inoculation of rhizobial isolates and PGPR isolates could be attributed to plant growth promoting abilities of PGPR isolates and; nitrogen fixing and PGP traits of rhizobial isolates. Selected rhizobial isolates RD20–3 (R1) and RK3–1 (R2) being GA producer and ammonia producer, stimulate plant growth through plant growth hormone production and protecting plants against phytopathogens. Both isolates were also capable of utilizing highest number of carbohydrates which depicts their ability to proliferate in diverse soils having varying carbon sources. Selected PGPR isolates NAG-K3 (P1) and CRC-J2 (P2) being IAA, GA producer and P solubilizer; and IAA, GA producer, Zn solubilizer and HCN producer, respectively improved plant growth through beneficial effects on seed germination, differentiation, proliferations and division of root cells and through biocontrol activity against harmful fungus. Phosphorus solubilizing capability of isolates increased the availability of phosphorus and through Zn solubilizing ability improves Zn nutrition of plants. Higher root growth and phosphorus availability renders the plant to uptake more amount of other nutrients also from soil and also increase the nitrogen fixing potential of rhizobial isolates which makes more amount of atmospheric nitrogen available to the plants to increase their vegetative growth, chlorophyll content, no. of trifoliate leaves, leaf area, root acid phosphatase activity and nitrogen and phosphorus status in plants. Higher root proliferation due to more differentiation and division of root cells increases the surface area which resulted higher water and nutrient absorption from soil. Higher uptake of nitrogen and phosphorus improved the starch synthesis and carbohydrate portioning in plant, resulted highest root and shoot length; dry biomass of root and shoot; and shoot: root ratio; when rhizobial (R1) and PGPR (P2) isolates applied conjointly at 100 kg N ha^{-1} application. N being the chief constituent of amino acids and protein; and P as constituent of phospholipid, both are the building blocks of protoplasm which forms the body structure of plant. Thus N and P improves the body structure and increase number of trifoliate leaves and leaf area of plants and various physiological parameters vis-à-vis dry biomass accumulation in plants. There are two types of phosphatases on the basis of pH as acid and alkaline phosphatase, because maximum activity occurs at low pH (6.5) and high range of pH (11.0), respectively. Acid level phosphatases secretion can vary with crop sp. [106]. Thus there is differential interspecific genetic variation in root enzyme secretion and acid phosphatase activity [107]. Co-inoculation of rhizobial isolates and PGPR isolates significantly improved root acid phosphatase activity in plants converting organic phosphorus compounds to inorganic-P. Concurrent exudation of organic acids and phosphatase by phosphate solubilizing microorganisms could enhance P solubility, by releasing bound organic phosphates and its mineralization by escalating the rate of hydrolytic cleavage [108].

As the main constituent of chlorophyll and an element of porphyrin ring, N content supplied by symbiotic nitrogen fixation and external application significantly improved the synthesis of chlorophyll 'a', chlorophyll 'b', chlorophyll 'a':'b' ratio in plant, when rhizobial and PGPR isolates applied in combination. Chlorophyll 'a' is always greater than chlorophyll 'b' and chlorophyll 'a' plays an important role in photosynthesis through absorbing light energy and converting it into chemical energy while chlorophyll 'b' as an accessory pigment absorb more light energy and transfers it to chlorophyll 'a' for photosynthesis. The chlorophyll 'a': chlorophyll 'b' ratio could be a useful indicator of N partioning within a leaf because this ratio is positively correlated with the ratio of PSII cores to light harvesting chlorophyll-protein complex.

Nitrogen and phosphorus uptake was significantly improved in plants could be attributed to higher N availability through symbiotic nitrogen fixation and high P availability through root acid phosphatase activity and phosphorus solubilizing ability of PGPR isolates. Significantly higher amount of fixed nitrogen in shoots, roots and grains of plant might be attributed to more nodulation and higher root growth due to phosphorus and nitrogen supplementation. The higher amount of nitrogen fixed in French bean due to combined inoculation of rhizobial isolate RD20–3 and PGPR isolate NAG-K3 over solitary inoculation of individual isolate is attributed to the higher phosphorus availability made due to P-solubilization and root acid phosphatase activity. The enhanced P-availability facilitates the more ATP synthesis which is required as a source of energy for carrying out the N_2-fixation by an enzyme nitrogenase. Number of pods plant^{-1}, pod yield, grains pod^{-1} and grain yield was significantly increased due to inoculation of French bean with rhizobial isolates and PGPR conjointly at 100 kg N ha^{-1} might be attributed to higher content of nitrogen and phosphorus in plant body which help in pod formation and grain formation in plants. Nitrogen as a chief constituent of protein and phosphorus also helps in protein synthesis, resulted highest protein content and protein yield plants receiving rhizobial and PGPR isolates both, at 100 kg N ha^{-1} application. Co-inoculation of rhizobial isolate RK3–1 (R2) and PGPR isolate CRC-J2 (P2) significantly reduced all plant growth parameters, physiological parameters, quality parameters, grain yield and nutrient uptake in plants compared to application of RK3–1 (R2) and CRC-J2 (P2), alone [69] which depicts the non-synergistic interaction between these rhizobial and PGPR isolates. Thus these two isolates RK3–1 (R2) and CRC-J2 (P2) may not be used as consortium to improve plant growth.

Co-inoculation of *Rhizobium* and PGPR showed a better nodulation which resulted higher shoot dry biomass and seed yield production. Beneficial effects of PGPR on symbiotic efficiency of rhizobia nodulating legume crops have also been reported by various scientists [64, 109]. According to Samavat et al. [110] the significant correlation has been observed between nitrogen absorption and improvement in growth of plant roots and shoots, as a result of *Rhizobium* and *Pseudomonas* interaction. Lucas Guarcia et al. [111] reported that the co-application of rhizobacteria and *Rhizobium* might have effects on their symbiotic relation with the host legume, depending on the applied isolates. Co-application of *Rhizobium* and *Pseudomonas* improved nodulation, leaf chlorophyll content and other growth factors under greenhouse conditions [110]. The increased chlorophyll content in plant leaves as the result of bacterial isolates co-inoculation could be due to the increased plant nutrition and photosynthesis [112]. An application of PGPR together with *Rhizobium* improved the growth and seed production of beans [62]. Mishra et al. [113] reported that combined application of *Rhizobium* + PSB + PGPR improved plant height and number of pods significantly. An over view of the combined dual inoculation of *A. chlorophenolicus* and *Enterobacter;* and triple inoculation of strain *B. megaterium*, *A. chlorophenolicus* and *Enterobacter* gave significantly higher plant height.

14. Conclusion

This study revealed the presence of efficient multi-trait rhizobial and PGPR isolates in French bean rhizosphere. Those rhizobial isolates which nodulate the plant under controlled conditions may be authenticated as *Rhizobium* and, having plant growth promoting traits may increase plant height, nodule number, dry biomass, chlorophyll content, grain yield, nutrient uptake and protein yield after inoculation. PGPR isolates from French bean rhizosphere possessed the efficient pant growth promoting traits *viz;* IAA production, GA production, P solubulization, Zn Solubilization, HCN production, ammonia production and siderophore production may be considered effective for improving plant growth, physiological and quality parameters after inoculation in plants. Inoculation of *Rhizobium* with higher levels of N application may have a positive influence on plant height, dry biomass production, leaf area content, chlorophyll content, grain yield, protein content and water soluble carbohydrate content whereas negative effect on nodule number and dry biomass of nodule. Co-inoculation of *Rhizobium* and PGPR improved the plant growth, physiological and quality attributes of French bean. Microbial consortium comprised of *Rhizobium* and PGPR may not only enhance growth and yield of French bean but may also reduce inorganic fertilizer application.

Conflict of interest

None.

Author details

Saroj Kumar Yadav[1*] and Kiran P. Raverkar[2]

1 Dr. Rajendra Prasad Central Agricultural University, Pusa, Samastipur, Bihar, India

2 G.B. Pant University of Agriculture and Technology, Pantnagar, Uttarakhand, India

*Address all correspondence to: sarojkumaryadav01@gmail.com

References

[1] Dubey R C, Maheshwari D K, Kumar H, Choure K. 2010. Assessment of diversity and plant growth promoting attributes of rhizobia isolated from *Cajanus cajan* L. Afr J. Biotech. 9: 8619-8629.

[2] Raverkar K P. 2017. To exploit the microbial biodiversity in various agro-ecologies for biofertilizer application in diverse cropping systems. *Progress Report.* All India Network Project on Soil biodiversity: Biofertil. 1-12.

[3] Sharma A, Johri B N, Sharma A K, Glick B R. 2003. Plant growth-promoting bacterium *Pseudomonas* sp. strain GRP3 influences iron acquisition in mung bean (*Vigna radiata* L. Wilzeck). Soil Biol. Biochem. 35: 887-894.

[4] Karthik C, Oves M, Sathya K, Padikasan I A. 2017. Isolation and characterization of multi-potential *Rhizobium* strain ND2 and its plant growth promoting activities under Cr (VI) stress. Arch. Agron. Soil Sci. DOI: 10.1080/03650340.2016.1261116

[5] Tsai S M, Bonetti R, Agbala S M, Rossetto R. 1993. Minimizing the effect of mineral on biological nitrogen fixation in common bean by increasing nutrient levels. Plant Soil 152:131-138.

[6] Kaur J, Khanna K, Kumari P, Sharma R. 2015. Influence of psychrotolerant plant growth-promoting rhizobacteria as co-inoculants with *Rhizobium* on growth parameters and yield of lentil (*Lens culinaris* Medikus). Afric. J. Microbiol. Res. 9: 258-264.

[7] Egli D B. 1998. Seed biology and the yield of grain crops. Wallingford, UK: CAB International. 32pp.

[8] Harlan J R. 1995. The Living Fields: Our Agricultural Heritage. Cambridge: Cambridge University Press. 271 pp.

[9] Heiser C B J. 1973. Seed to civilization-The story of Man's Food. San Francisco: W.H. Freeman. 243 pp.

[10] Jensen E S, Nielsen H. 2003. How can increased use of biological N_2-fixation in agriculture benefit the environment? Plant Soil 252: 177-186.

[11] Debouck D.1991. Systematics and Morphology. In*:* A. van Schoonhoven and O. Voysest, (eds). Common beans research for crop improvement. Cali, Colombia: CAB International, CIAT.

[12] Broughton W J, Hernz G, Blair M, Vanderleyden J. 2003. French beans (*Phaseolus vulgaris* L.) model food legumes. Plant soil 252: 55-128.

[13] Anonymous 2005. Food and Agriculture Organization of the United Nations. FAO STAT database.*http:/www.fao, org.*

[14] Nleya T M, Slinkard A E, Vandenberg A. 2001. Differential performance of pinto beans under varying levels of soil moisture. Canad. J. Plant Sci. 81: 233-239.

[15] Maesen L J G, Somaatmadja S. 1989. *Phaseolus vulgaris* L. In*:* L. J. G. van der Maesen and S. Somaatmadja, (eds). Plant Resources of South-East Asia 1.Pulses. Wageningen: Pudoc/ Prosea, 60-63.

[16] Silbernagel M J, Janssen W, Davis J H C. 1991. Snap bean production in the tropics: implications for genetic improvement. In*:* A. van Schoonhoven and O. Voysest, (eds). Common Beans: Research for Crop Improvement. Wallingford, UK and Cali, Columbia, 835-862.

[17] Hardarson G, Atkins C. 2003. Optimising biological N_2-fixation by legumes in farming system. Plant Soil 252: 41-54.

[18] Mellor R M, Werner D. 1990. Legume nodule biochemistry and function. In: P. M. Gresshof, (ed). Molecular Biology of Symbiotic Nitrogen Fixation. Boca Raton: CRC Press, Inc. 111-129.

[19] Dupont L, Alloing G, Hopkins J, Hérouart D, Frendo P. 2012. The legume root nodule: From symbiotic nitrogen fixation to senescence. Senes. 1-34.

[20] Somasegaran P, Hoben H J. 1994. Handbook for rhizobia.Methods in legume-*Rhizobium* technology. Springer Verlag, New York.

[21] Howieson J, Ballard R. 2004. Optimizing the legume symbiosis in stressful and competitive environments within southern Australia- some contemporary thoughts. Soil Biol. Biochem. 36: 1261-1273.

[22] O' Hara G W, Howieson J G, Graham P H. 2003. Nitrogen fixation and agricultural practice. In: G. J. Leigh, (ed). Nitrogen Fixation at the Millennium. Amsterdam: 391-420.

[23] Graham P H, Vance C P. 2000. Nitrogen fixation in perspective: an overview of research and extension needs. Field Crop Res. 65: 93-106.

[24] Jordan D C. 1984. Family II. *Rhizobiaceae.* In Bergey's manual of systematic bacteriology. Vol. I (eds. By N. R.Krieg and J. G. Holt Williams and Wilkins Co., Baltimore, M. D). pp. 232-242.

[25] Martinez-Romero E, Segovia L, Mercante F M, Franco A A, Graham P H, Prado M A. 1991. *Rhizobium tropici*, a novel species nodulating *Phaseolusvulgaris* beans and *Leucaena sp*. trees. Int. J. Syst. Bacteriol. 41: 417-426.

[26] Aguilar O M, Riva O, Peltzer E. 2004. Analysis of *Rhizobium etli* and of its symbiosis with wild *Phaseolus vulgaris* supports co-evolution in centers of host diversification. Proc. Nat. Acad. Sci. USA 101: 13548-13553.

[27] Vincent J M. 1970. Manual for practical study of root nodule bacteria.IBP Hand Book '5, Blackwell Scientific Publishing Company, Oxford.

[28] Hungria M, Andrade D S, Chueire L M d O, Probanza A, Guttierrer-Manero F J, Megias M. 2000. Isolation and characterization of new efficient and competitive bean (*Phaseolus vulgaris* L.) rhizobia from Brazil. Soil. Biol. Biochem. 32: 1515-1528.

[29] Yadav S K, Raverkar K P, Chandra R, Pareek N, Chandra S. 2019. Isolation, authentication and evaluation of rhizobial isolates from the soils of North-West Himalayas in French bean (*Phaseolus vulgaris* L.). Int. J. Curr. Microbiol. Appl. Sci. 7:141-149.

[30] Koskey G, Simon W, Ezekiel M. 2018. Genetic characterization and diversity of *Rhizobium* isolated from root nodules of Mid-Altitude climbing Bean (*Phaseolus vulgaris* L.) varieties. Front. Microbiol. 9:1-10.

[31] Rai R, Sen A. 2015. Biochemical characterization of French bean associated rhizobia found in North Bengal and Sikkim. J. Acad. Indu. Res. 4: 10-18.

[32] Holt J G, Krieg N R, Sneath P H A, Staley J T, Williams S T. 1994. In Bergey's Manual of determinative bacteriology. Williams and Wilkins Press, Baltimore. USA

[33] Deshwal V K, Chaubey A. 2014. Isolation and characterization of *Rhizobium leguminosarum* from root nodule of *Pisum sativum* L. J. Acad. Indus. Res. 2: 464-467.

[34] Deka A K, Azad P. 2006. Isolation of *Rhizobium* strains: cultural and biochemical characteristics. Legume Res. 29: 209-212.

[35] Singh B, Kaur R, Singh K. 2008. Characterization of *Rhizobium* strain isolated from the roots of *Trigonellafoenumgraecum* (fenugreek). Afr. J. Biotechnol. 7: 3671-3676.

[36] Erum S, Bano A. 2008. Variation in phytohormone production in *Rhizobium* strains at different altitudes of Northern areas of Pakistan. Int. J. Agric. Biol. 10: 536-540.

[37] Gauri A, Bhatt R, Pant S, Bedi M, Naglot A. 2011. Characterization of *Rhizobium* isolated from root nodules of *Trifolium alexandrinum*. J. Agric. Technol. 7: 1705-1723.

[38] Hunter W J, Kuykendall L D, Manter D K. 2007. *Rhizobuim selenireducens* sp. nov: A selenite reducing- proteobacteria isolated from a bioreactor. Curr. Microbiol. 55: 455-460.

[39] De Oliveira A N, de Oliveira L A, Andrade J S, Chagas J A F. 2007. *Rhizobia* amylase production using various starchy substances as carbon substrates. Braz. J. Microbiol. 38: 208-216.

[40] Aneja K R. 1996. Experiments in Microbiology, Plant Pathology, Tissue Culture and Mushroom Cultivation. 2nd edition, New Age International Publishers, New Delhi, India. pp. 240-249.

[41] Kumari B S, Ram M R, Mallaiah K V. 2010. Studies on nodulation, biochemical analysis and protein profiles of *Rhizobium*. Malay J. Microbiol. 6: 133-139.

[42] Kloepper J W. 1994. Plant growth promoting rhizobacteria. In: Okon, Y. (Ed.), Azospirillum / Plant Associations. CRC Press, Boca Raton, FL, USA, pp. 111-118.

[43] Anitha G, Kumudini B S. 2012. Isolation and characterization of fluorescent *Pseudomonads* and their effect on plant growth promotion. J. Environ. Biol. 29: 627-635.

[44] Battu P R, Reddy M S. 2009. Isolation of secondary metabolites from *Pseudomonas fluorescens* and its characterization. Asian J. Res. Chem. 2: 26-29.

[45] Rodríguez-Cáceres E A. 1982. Improved medium for isolation of *Azospirillum spp*. App. Environ. Microb. 44: 990-991.

[46] Garcia de Salamone I E, Döbereiner J, Urquiaga S,Boddey R M. 1996. Biological nitrogen fixation in *Azospirillum* strain maize genotype associations as evaluated by ^{15}N isotope dilution technique. Biol. Fertil. Soils 23: 249-256.

[47] Maingi J, Shisanya C, Gitonga M N, Hornetz B. 2001. Nitrogen fixation by common bean (*Phaseolus vulgaris* L.) in pure and mixed stands in semi-arid south-east Kenya. Eur. J. Agron. 14:1-12.

[48] Anglade J, Billen G, Garnier J. 2015. Relationship for estimating N_2-fixation in legumes: Incidence for N balances of legume-based cropping systems in Europe. Ecosph. 6: 1-24.

[49] Muthini M, Maingi J M, Muoma J O, Amoding A, Mukaminega D, Osoro N. 2014. Morphological assessment and effectiveness of indigenous rhizobia isolates that nodulate *P. vulgaris* in water hyacinth compost testing field in Lake Victoria basin. Ecosph. 4: 718-738.

[50] Bala A, Abaidoo R, Woomer P. 2013. Strain isolation and characterization protocol *Rhizobia*. Annal Microbiol. 64: 209-218.

[51] Pohajda I, Huić Babić K, Rajnović I, Kajić S, Sikora S. 2016. Genetic diversity and symbiotic efficiency of indigenous common bean rhizobia in Croatia. Appl. Environ. Microbiol. 54: 468-474.

[52] Musandu A A O, Ogendo O J. 2001. Response of common bean to *Rhizobium* inoculation and fertilizers. J. Food Tech. 6: 121-125.

[53] Asadi-Rahmani H, Rasanen L A, Afshari M, Lindstrom K. 2011. Genetic diversity and symbiotic effectiveness of rhizobia isolated from root nodules of *Phaseolus vulgaris* grown in soil of Iran. Appl. Soil. Ecol. 48: 287-293.

[54] Deshwal V K, Dubey R C, Maheshwari D K. 2003. Isolation of plant growth promoting strains of *Bradyrhizobium* (*Arachis* sp.) with biocontrol potential against *Macrophomina phaseolina* causing charcoal rot of peanut. Curr. Sci. 84: 443-448.

[55] Asadi R H, Afshari M, Khavazi K, Nourgholipour F, Otadi A. 2005. Effects of common bean nodulating *rhizobia* native to Iranian soils on the yield and quality of bean. Iranian J. Soil Water Sci. 19: 215-225

[56] Mnasri B, Elarbi Aouani M, Mhamdi R. 2007. Nodulation and growth of common bean (*Phaseolus vulgaris*) under water deficiency. Soil Biol Biochem. 39: 1744-1750.

[57] Ndlovu T J. 2015. Effect of *Rhizobium phaseoli* inoculation and phosphorus application on nodulation, growth and yield components of two Dry bean (*Phaseolus vulgaris*) cultivars. Mini dissertation Submitted in partial fulfilment of the requirements of the degree of Master of Science in Agriculture (Agronomy) at the University of Limpopo, South Africa.

[58] Rodriguez A, Frioni L. 2003. Characterization of rhizobia causing nodules on leguminous trees native to Uruguay using the rep-PCR technique. Rev. Argent. Microbiol. 35: 193-197.

[59] Das K. 2017. Varietal performance of bush type French bean varieties (*Phaseolus vulgaris* L.) for growth, fresh pod yield and quality. Thesis Submitted to the Uttar Banga Krishi Viswavidyalaya In partial fulfillment of the requirements for the Degree of Master of Science.

[60] Meena J K, Ram R B, Meena M L. 2018a. Studies on bio-fertilizers yield and quality traits of French bean (*Phaseolus vulgaris* L.) cultivars under Lucknow condition. J. Pharmaco. Phytochem. 7: 1571-1574.

[61] Togay N, Togay Y, Mesut K, Turan M. 2008. Effects of rhizobium inoculation, sulfur and phosphorus applications on yield, yield components and nutrient uptakes in chickpea (*Cicer arietinum* L.). Afric. J. Biotechnol. 7: 776-782.

[62] Yadegari M, Rahmani H A. 2010. Evaluation of bean (*Phaseolus vulgaris)* seeds inoculation with *Rhizobium phaseoli* and plant growth promoting rhizobacteria on yield and yield components. African J. Agril. Res. 5: 792-799.

[63] Arruda N B D, Dehereiner J, German C N. 1968. Inoculation, N manuring and time pelleting with three soybean varieties. Pesq. Agropec. Brasil. 3: 201-204.

[64] Figueiredo M V B, Burity H A, Martinez C R, Chanway C P. 2008. Alleviation of water stress effects in common bean (*Phaseolus vulgaris* L.) by co-inoculation *Paenibacillus x Rhizobium tropici*. Appl. Soil Ecol. 40: 182-188.

[65] Otieno P E, Muthomi J W, Chemining'wa N G, John H N. 2009. Effect of rhizobia inoculation, farmyard manure and nitrogen fertilizer on nodulation and yield of food grain legumes. J. Biol. Sci. 9: 326-332.

[66] Bambara S, Ndakidemi P A. 2010. Effect of *Rhizobium* inoculation, lime and molybdenum on nitrogen fixation

of nodulated *Phaseolus vulgaris* L. African. J Microbiol. Res. 4: 682-696.

[67] Namvar A, Seyed Sharifi R, Sedghi M, Asghari Zakaria R, Khandan T, Eskandarpour B. 2011. Study on the effects of organic and inorganic nitrogen fertilizer on yield, yield components and nodulation state of chickpea (*Cicer arietinum* L.). Commun. in Soil Sci. Plant Anal. 42: 1097-1109.

[68] Meena J K, Ram R B, Meena M L. 2018b. Efficacy of Bio-fertilizers on vegetative characters of French Bean (*Phaseolus vulgaris* L.) Cultivars. Int J Pure App Biosci. 6:1351-1355.

[69] Yadav S K. 2019. Development of microbial consortium for enhancing French bean (*Phaseolus vulgaris* L.) productivity. Thesis submitted to G. B Pant University of Agriculture and Technology, Pantnagar, Uttarakhand, India.

[70] Aminifard M H, Aroiee H, Karimpour S. 2010. Response of egg plant (*Solanum melongena* L.) to different rates of nitrogen under field condition. J. Cent. Eur. Agri. 11: 453-458.

[71] Salvagiotti F, Daniel T, Dobermann A. 2008. Growth and nitrogen fixation in high-yielding soybean: impact of nitrogen fertilization. Agron. J. 101:958-970.

[72] Negi S C, Shekhar J. 1993. Response of French bean (*Phaseolus vulgaris*) genotypes to nitrogen. Indian J. Agron. 38: 321-322.

[73] Dwivedi D K, Singh H, Singh K M, Shahi B, Rai J N. 1994. Response of French bean *(Phaseolus vulgaris)* to population density and nitrogen levels under mid-upland situation in north-east alluvial plains of Bihar. Indian J. Agron. 39: 581-583.

[74] Kushwaha B L. 1994. Response of French bean *(Phaseolus vulgaris L.)* to nitrogen application in north Indian plains. Indian J. Agron. 39: 34-37.

[75] Saxena K K, Verma V S. 1995. Effect of nitrogen, phosphorus and potassium on the growth and yield of French bean (*Phaseolus vulgaris L.*). Indian J. Agron. 40: 249-252.

[76] Singh D P, Rajput A L. 1995. Effect of spacing and nitrogen on yield and economics of French bean. Haryana J. Hort. Sci. 11: 122-127.

[77] Singh A K, Singh S S. 2000. Effect of planting date, nitrogen and phosphorus levels on yield contributing characters in French bean. Legume Res. 23: 33-36.

[78] Dhanjal R, Prakash O, Ahlawat I P S. 2001. Response of French bean *(Phaseolus vulgaris)* varieties to plant density and nitrogen application. Indian J. Agron. 46:277-281.

[79] Prajapati M P, Patil L R, Patel B M. 2003. Effect of integrated weed management and nitrogen levels on weeds and productivity of French bean *(Phaseolus vulgaris L.)* under north Gujarat conditions. Legume Res. 26: 77-84.

[80] Veeresh N K. 2003. Response of French bean (*Phaseolus vulgaris L.*) to fertilizer levels in Northern Transitional Zone ofKarnataka. M.Sc. (Agri.) Thesis submitted to University of Agricultural Science, Dharwad (India).

[81] Jagdale R B, Khawale V S, Baviskar P K, Doshinge B B, Kore M S. 2005. Effect of inorganic and organic nutrients on growth and yield of French bean *(Phaseolus vulgaris* L.*)*. J Soil Crops 15: 401-405.

[82] Singh R, Singh Y, Singh O N, Sharma S N. 2006. Effect of nitrogen and micronutrients on growth, yield and nutrient uptake by French bean. Indian J. Pulse Res. 19: 67-69.

[83] Sharma H M, Singh R N P, Singh H, Sharma R P R. 1996. Effect of rates and timings of N application on growth and yield of winter rajmash. Indian J. Pulse Res. 9: 23-25.

[84] Rajput V, Acharya P, Singh G. 1999. Effect of dates of sowing and graded doses of nitrogen on growth and yield of french bean cv. Contender in eastern U.P. Orissa J. Hort. 27: 39-42

[85] Behura A K, Mahapatra P K, Swain D. 2006. Effect of irrigation and nitrogen on physiological growth parameters, yield and yield attributes of rajmash *(Phaseolus vulgaris L.)*. Res. on crops 7: 92-95.

[86] Escalante A, Rodriguez M T, Escalante E. 1993. Effect of nitrogen on development and abscission of reproductive organs of beans. Agron Mesoameric. 10: 47- 53.

[87] Cechin I, Fumis T F. 2004. Effect of nitrogen supply on growth and photosynthesis of sunflower plants grown in the greenhouse. Plant Sci. 166: 1379-1385.

[88] Hegde D M, Srinivas K. 1989. Effect of irrigation and nitrogen on growth, yield and water use of French bean. Indian J. Agron. 34: 180-184

[89] Rana N S, Singh R. 1998. Effect of nitrogen and phosphorus on growth and yield of French bean (*Phaseolus vulgaris L.*). Indian J. Agron. 43: 367-370.

[90] Namvar F, Mohamad R, Rahman H S. 2013. Antioxidant, antiproliferative and antiangiogenesis effects of polyphenol-rich seaweed (*Sargassum muticum*). Bio Med Res. Intern. 10:1- 10.

[91] Ralcewicz M, Knapowski T, Kozera W,Barczak B. 2009. Technological value of 'Zebra' spring wheat depending on the nitrogen and magnesium application method. J. Central Eur. Agri. 10: 223-232.

[92] Waraich E A, Ahmad R, Ashraf M Y. 2011. Role of mineral nutrition in alleviation of drought stress in plants. Australian J. Crop Sci. 5: 764-777.

[93] Ashraf M A, Asif M, Zaheer A, Malik A, Ali Q, Rasool M. 2001. Plant growth promoting rhizobacteria and sustainable agriculture. Afr. J. Microbiol. Res. 7: 704-709.

[94] Yoseph T, Shanko S. 2017. Growth, symbiotic and yield response of N-fertilized and *Rhizobium* inoculated common bean (*Phaseolus vulgaris* L.). Afric. J. Plant Sci. 11: 197-202.

[95] Omoregie A U, Okpefa G O. 1999. Effects of time of application of nitrogen on nodulation, dry matter and mineral nutrition of cowpea *Vigna unguiculata* L. (Walp) in the Delta area of Nigeria. Nigerian Agril J. 30: 32-40.

[96] Kucuk C, Kivanc M, Kinaci E. 2006. Characterization of *Rhizobium* sp. isolated from Bean. Turky J. Biol. 30: 127-132.

[97] Appunu C, Sen D, Singh M K, Dhar B. 2008. Variation in symbiotic performance of *Bradyrhizobium japonicum* strains and soybean cultivars under field conditions. J. Eur. Cent. Agr. 9: 185-190.

[98] Sajid M., Rab A, Hussain S A, Iqbal Z. 2011. Influence of rhizobium inoculation on growth and yield of groundnut cultivars. Sarhad J. Agric. 27: 573-576.

[99] Higa T. 1994. Effective microorganisms: A new dimension for nature Farming. p. 20-22. In: Parr JF, Hornick SB and Simpson ME (Eds), Proceedings of the Second International Conference on Kyusei Nature Farming. U. S. Department of Agriculture, Washington DC, USA.

[100] Halverson L J, Handelsman J. 1991. Enhancement of soybean nodulation by

Bacillus cereus UW85 in the field and in a growth chamber. Appl. Environ. Microbiol. 57: 2767-2770.

[101] Lifshitz R, Kloepper J W, Kozlowski M. 1987. Growth promotion of canola seedlings by a strain of *Pseudomonas putida* under gnobiotic condition. Can. J. Microbiol. 33: 390-395.

[102] Sessitsch A, Coenye T, Sturz AV, Glick B R, Nowak J. 2005. *Burkholderia phytofirmins* sp Nov., a novel plant-associated bacterium with plant beneficial properties. Int. J. Syst. Evol. Microbiol. 55: 1187-1192

[103] Karakurt H, Kotan R, Dadasoglu F, Sahin F. 2011. Effects of plant growth promoting rhizobacteria on fruit set, pomological and chemical characteristics, colour values and vegetative growth of sour cherry (*Prunus cerasus*). Turky J. Biol. 35: 283-291.

[104] Stefan M, Dunca Z, Olteanu L, Cojocaru D. 2010. Soybea (*Glycine max* L.) inoculation with *Bacillus pumilus* promotes plant growth and increases seed protein yield: relevance for environmentally-friendly agricultural applications. Carpathian J. Earth Environ. Sci. 5: 131-138.

[105] Arora N K., Tewari S, Singh R. 2013. Multifaceted plant-associated microbes and their mechanisms diminish the concept of direct and indirect PGPRs In: Arora NK (ed.) Plant Microbe Symbiosis: Fundamentals and Advances. Springer, 411-449.

[106] Nuruzzaman M, Lambers H, Michael D A, Veneklaas E J. 2006. Distribution of carboxylates and acid phosphatase and depletion of different phosphorus fractions in the rhizosphere of a cereal and three grain legumes. Plant Soil 281:109-120

[107] Tadano T, Ozawa K, Sakai H, Osaki M, Matsui H. 1993. Secretion of acid phosphatase by the roots of crop plants under phosphorus-deficient conditions and some properties of the enzyme secreted by lupin roots. Plant Soil 155: 95-98.

[108] Trolove S N, Hedley A D, Kirk C, Loganathan P. 2003. A progress in selected areas of rhizosphere research on P acquisition. Aust. J. Soil Res. 41: 471-499.

[109] Dashti N, Zhang F, Hynes R, Smith D I. 1997. Application of plant growth-promoting rhizobacteria to soybean (*Glycine max* L.) increases protein and dry matter yield under short-season conditions. Plant Soil 188: 33-41.

[110] Samavat S, Ahmadzadeh M, Behboudi K, Besharati H. 2013. Comparing the Ability of *Rhizobium* and *Pseudomonas* isolates in controlling bean damping-offcaused by *Rhizoctonia solani* Kuhn. J. Biol. 3: 1-12.

[111] Lucas Guarcia J A, Probanza A, Ramos B, Barriuso J, Gutierrez M. 2004. Effects of inoculation with plant growth promoting rhizobacteria and *Sinorhizobium fredii* on biological nitrogen fixation, nodulation and growth of *Glycine max* cv. *Osumi*. Plant Soil 267: 143-153.

[112] Bashan Y, Harrison S K, Whitmoyer R E. 1990. Enhanced growth of wheat and soybean plants inoculated with *Azospirillum brasilense* is not necessarily due to general enhancement of mineral uptake. Appl. Environ. Microb.56: 769-775.

[113] Mishra P K, Mishra G, Selvakumar S C, Bisht J K, Kundu S, Gupta S. 2008. Characterization of psychrotolerant plant growth promoting *Pseudomonas* sp. Strain PGERs 17 (MTCC 9000) isolated from North Western Indian Himalayas. Ann Microbiol. 58: 561-568.

Chapter 6

Application of Allelopathy in Crop Production

Wasan S. Hussain and Mahmoud M. Abbas

Abstract

Need for food production has been increasing greatly in recent years throughout the world. The interest on the supply of quality of food has also increased, but a significant loss of crop production was observed annually, especially the main cereal crops, including rice, wheat and maize, due to the presence of weeds accompanying them in the growing season. Allelopathy has emerged as an alternative approach to solve problem agriculture that including: crop rotations, intercropping, crop residue incorporation and aqueous extracts all that used to explore allelopathy for pest management, enhancement of growth and crop production. As will allelopathic consider as weeds, insect and diseases natural control. Secondary metabolites biosynthesis of at high rates have a great role in provides defense against abiotic stresses. In plant rhizosphere allelochemicals exuded improve nutrient acquisition through the processes of solublization; biological nitrification; chelation and selected retention. In this chapter, application of the allelopathic phenomenon in crop production is discussed and his roller in managing agricultural pests and improving the productivity of agricultural systems. It was found that allelochemicals promote plant growth and production at low concentration; however it can suppress the growth if applied at high concentrations, for that can be used allelopathic compounds for weed control by used high concentrations of plant residues or aqueous extracts of plant.

Keywords: allelopathy, weed control, crop production, allelochemicals

1. Introduction

Productivity is a good indicator of the conditions suitable, since it reflects directly changes in the quality of soil. The main target of soil management for agriculture is to create favorable conditions for good crop growth, seed germination, plant development, and good harvest. Increased yield of crops associate with good field management practices through increase soil fertility, crop nutrition, weed control, insect and diseases management.

Population increasing in world is a threat to agricultural sustainability and food security for this reason we need to increase crops productive to solve multiple issues in modern agriculture by crop rotations, cover crops, intercropping, mulching, plant residue and aqueous extracts of plant pest management, stress mitigation, and growth enhancement in crop production all these can be due to allelopathic application.

Allelopathy is an important mechanism of plant by the release of plant-produced secondary metabolites or decomposition products of microbes to environment. Allelopathy plays a role main in natural ecosystems, that found many of mechanism to released allelochemicals into the soil rhizosphere that: volatilization, residues decomposition, and root exudation and leaching [1]. Allelochemicals affects is an inhibitory or stimulatory of several roles ecological of these chemicals compound, including plant defense, nutrient chelation, and regulation of soil microorganism, all that depend to concentration of that allelochemicals [2].

In more studies found that allelochemical aqueous extracts application at lower concentrations stimulates germination and growth of many crops and enhancement productivity of crop [3, 4].

Role inhibitory of allelochemicals is well explored that previously was only dimension are known from allelopathy. Directly and indirectly role for allelopathy has been used for weed management by biological control [5]. Aslam et al. [6] presented a contemporary synthesis of the existing data that how allelopathy can be exploited: (a) to biologically control of insects, and disease, (b) enhance of quality of soil by adding decomposition of crop plants residues as nutrients and improvement soil as microbes environment, (c) crop diversification increase by reducing the weeds and pests infestation, (d) to develop biological pesticides from crop plants with a novel mode of action, (e) confer abiotic stresses tolerance (**Table 1**).

Impartment of this study lies in use of allelopathic compounds as a bio stimulants for growth at low concentration, and biological control at high concentration.

Study aims to use allelopathic phenomena in:

1. Biologically control of insects, disease and **weeds**.

2. Enhance of quality of soil by adding decomposition of crop plants residues as nutrients and improvement soil as microbes environment.

3. Increased yield of crops through increase soil fertility, crop nutrition, weed control, insect and diseases management.

Allelopathic source	Insect-pest/pathogen suppressed	Percent control	Reference
Neem (*Azadirachta indica*)	Corcyra cephalonica	26.0	Batish et al. [7]
Tomato (*Lycopersicon esculentum*)	Flower thrip (*Taeniothrips sjostedti* Trybom)	32.0	Batish et al. [7]
Hot pepper (*Capsicum unnuum*)	Pod borer (*Heliothis armigera* Hb.)	54.0	Batish et al. [7]

Table 1.
Insect-pests and disease control through allelopathy.

2. Allelopathy phenomenon and its effect on agricultural production

Through a comprehensive review of many research carried out in the field of allelopathy, we notice that it has doubled several times, by physiologists, plant, soil, weeds and natural product chemists.

Studies mainly on allelopathy involve study extracts effect obtained from various plant parts (stem, root, seeds, and leaves) appear them be the most important source of phytotoxicity compounds in plants [8].

Studies have indicated the presence of many crops that demonstrated the allelopathic effort in other crops that accompany them in the field or track them in agriculture, through the release of allelopathic compounds into the environment through leaching and root excudates, and the decomposition of plant residues by microbiology decomposition as well as by volatilization method [9].

These compounds have inhibitory and stimulating effects on plants and microorganisms and through their effect on many vital activities, noting that the effects of allelopathic compounds depend on their nature and concentration, as some compounds cause inhibitory effects on seed germination and growth, while other compounds cause stimulating effects.

The results of [10] showed that the plant residues of the sunflower *Helianthus annuus* L. within different growth stages (seedling, elongation, flowering, and maturity) caused a reduction in the germination and growth of wheat varieties *Triticum aestivum* L. and two varieties *Helianthus annuus* L., and that the flowering stage had a clear allelopathic effect of the tested species compared with the rest of the stages.

In another study, the two researchers indicated that adding tomato and rice residues to the soil caused inhibition of growth and some of the yield traits of varieties of wheat growing in soils added to rice residues, while tomato residues caused an increase in these traits.

In crop mixtures, amount of allelochemicals released effects by ecosystems depend on the environmental factors (temperature, water and level of soil nutrient), planting practices and crop species [11].

Continued emergence of additional information about the mechanisms of influence of allelopathic compounds in terms of selectivity, secretion, persistence, and genetic regulation mechanisms pose a constant challenge to plant scientists with the need to develop modern strategies that would enhance the protection of biological diversity [12].

2.1 Roles of allelopathy in plant biodiversity and sustainable agriculture

Released allelopathic compounds from plants that might be detrimental or beneficial to the growth of receptor plants. These compounds are involved in the environmental complex of managed or natural ecosystems.

Important role play that allelopathic compounds in the plant diversity determination, dominance, succession, and natural vegetation and productivity of plant in the agroecosystems. One of impartment reason decrease of crop productivity overuse of synthetic herbicide often environmental hazards, nutrient deficiency, an imbalance of soil microorganisms, and causes change in properties of soil physicochemical [13].

Plant residues released phenolic compounds present in soil are considered to be one of the main factors to allelopathic effects for plants growing .

Advantageous of Intercropping systems are they can provide higher crop yield and diversity with fewer related to pests and weeds than monocultures, systems plant interactions in between crops and weeds and intercropped crop species are still not well understood.

Modern agriculture is dominated by intensive monocultures, which causes the loss of biodiversity and ecological functions [11]. Biodiversity of an ecosystem maintains ecological services such as nutrient cycling, tolerance to pest occurrences and disease outbreaks [14].

Diversification of Agricultural contains agricultural practices that use principles of ecological to increase the productivity and agroecosystems [15]. Number of studies on diversification of agricultural such as agroforestry and intercropping (use of two or more crop species simultaneously) that is type of sustainable agriculture [16].

Positive interactions among plants are defined as the improvement processes of plants to harsh environments and increase resource availabilities to species [17]. If selected well intercropped species can be promoted facilitative interactions [18]. The common types In intercropping systems are mixture are legumes and non-legume species combination due to the capacity of many species of family Leguminosae to biological nitrogen fixation (BNF) in the [19].

It is found that intercropping legumes the microbial communities can alter around the rhizosphere., the bacterial community structure is different in the rhizosphere of faba bean and wheat intercropping from that in wheat monoculture [20]. Microorganisms activities can change the availability of soil nutrient and benefit plants. He et al. [21] have found that rhizobial communities are changes by intercropping maize with chickpea and soybean.

Alrababah et al. [22] found that interactions between crops and trees is necessary for producing crops and conserving forests especially within the threatened Mediterranean forest ecosystems, study explored the allelopathic effects of green and senescent leaf and soil extracts of two agroforestry trees Pinus and Quercus on the germination of wheat, barley, lentil, chickpea, and fababean as the major grain crops of Jordan. Results showed reduced seed germination of all crops.

2.2 Allelopathy and weed management

Important role can be allelopathy plays in agro-ecosystems that leading to a wide range of influences and interactions in a biotic communities. Allelopathy can be effects on plants including microorganisms direct or indirect effect, by released of natural products in to environment [1].

Crop allelopathic properties can be make one species more persistent to a native species. Therefore, potentially of these crops may be harmful to both naturalized as well as agricultural settings. While on other side, allelopathic crops provide strong potential for the development of cultivars that are more highly weed suppressive in managed settings [23].

In agro ecosystems, weeds compete with crop plants for resources, which reduce crop yield and deteriorate their quality, The most effective method of weed suppression is biological control which is a natural process, environment-friendly, low cost, and with high public acceptance. Biological methods include the use of insects, fungus, bacteria, allelopathic crops, cover crops, and mulching. Legume crop such as velvet bean can reduce weed biomass by 68% [24].

Allelopathi is a biological phenomenon that affects neighboring plants or that is followed it in agriculture through the production of chemical compounds which is one of the main factors limiting the growth of plants, can benefit from this phenomenon in reducing the growth of the weeds using crops have Allopathic effect, the studies found many crops that possess the effect of allopathic (wheat, sunflower, sorghum, barley, corn) [25].

Use of chemical control is one of the economically costly methods in addition to causing pollution in the environment, that affecting human health and the ecosystem, the continuous use of chemical herbicides creates generations of weeds resistant to those herbicides [26] for that studies have tended to find alternatives that are less expensive and environmentally friendly, as well as not affecting the genetic makeup of weeds. That alternative is the biologic control using the allelopathic compounds released from different parts of the plant, which can inhibit and reduce the growth of some weeds [27]. Biological control of the weeds using allelopathic compounds as herbicides has been widely used as a safe, useful and less costly method.

2.3 Allelopathy and insect management

Allelopathy is a naturally occurring ecological phenomenon of interference among organisms that may be employed for managing weeds, insect pests and diseases in field crops. In field crops, However, according to [1] allelopathy is the influence of one plant on the growth of another one, including microorganisms, by the release of chemical compounds into the environment. These chemicals are usually secondary plant metabolites or by products of the principal metabolic pathways in plants.

Many plants have natural defense mechanism against insect pests. They utilize secondary metabolites for this purpose. Neem (*Azadirachta indica*) produces allelochemicals, azadrachtin, salannin and nimbin [28].

3. Allelopathy promotion plant growth

Plants Growth promotion by other plants, as well as that of microorganism by plants and other microorganisms, is discussed. Agro system in mixed culture with wheat enhances growth and yield of wheat. Soil amended with shoots of *Solanum nigrum*, enhances the soybean growth and nodulation. Growth and yield of several legumes are enhanced by mixed culture with *Heliotropium peruvianum* [29].

The use of biostimulants, which are defined as substances or materials other than nutrients and pesticides that can be used to regulate the physiological processes in plants to stimulate their growth, Biostimulants promote plant growth and development throughout the crop's life cycle, from the seed stage to mature plants, by improving metabolic efficiency, resulting in increased yield and enhanced crop quality, and facilitating nutrient assimilation, translocation, increasing nutrient use efficiency continues to be a major challenge for world agriculture [30].

Thereby increasing plant tolerance to and recovery from abiotic stresses, allelochemicals are among biostimulants.

In laboratory and field has been reported to inconsistency between plant growth-promoting rhizobacteria effectiveness "PGPR" that stimulate growth of plant and yield.

In field this inconsistency results from PGPR applications can be solved by improved knowledge of interplay between host and introduced PGPR inoculant in rhizosphere under field conditions to reduced chemical fertilizer quantity can application of biofertilizer for maintaining threshold levels of crop productivity [31].

(PGPR) may be encourage plant growth by producing growth regulators; facilitating of nutrient uptake; mineralization accelerating; plant stressreducing; nodulationstimulating, providing nitrogen fixation; mycorrhizal promoting; suppressing plant diseases; and functioning as nematicides and insecticides. There are many PGPR are fluorescent pseudomonads (*Pseudomonas fluorescens*); and other bacteriaare known as well as (Bacillus sp., Azotobacter sp., Acetobacter sp., Azospirillum sp.).

Allelochemicals released by plants have promotory effects at low concentrations [32, 33]. Previous studies have elucidated the positive role of secondary metabolites, hormones and some other natural compounds produced by plants, in plant growth promotion [34].

Allelopathic effect in nutritional and ecological relevance in the soil system.

Relationship between soil characteristics and allelochemicals at two ways

The level of phytotoxicity is affected in soil characteristic, and they are closely linked to each other and effect on retention, transport and transformation processes of allelochemicals in soil. We need to understand the interactions involved in soil allelopathy and to create new opportunities for a sustainable control of agroecosystems.

Allelopathy has offered a new alternative for the development of eco-friendly agricultural practices, with the dual purpose of enhancing crop productivity and maintaining ecosystem stability [35]. Allelopathy involves the positive or negative effects of a plant (donor), including microorganisms, on neighboring plants (targets) through the release of chemical compounds into the environment by (leaching, volatilization, root exudation, decomposition) **Figure 1**, mostly in the soil. According to [36], it is possible to distinguish between direct plant-plant allelopathic interference (allelopathy in the narrow sense) and indirect allelopathy.

Influence of meteorological, soil and plant factors on the phytotoxicity of allelochemicals in soil [37]. Meteorological and plant factors can be effect in quantity and quality of allelochemicals released from plant (donor). When released to soil system, several of soil factors may be influence retention; transport and transformation processes of allelochemicals in soil and their presence in soil solution in order to be absorbed by other plant (target).

Allelochemicals effects on abiotic and biotic soil processes can effect in other plants. Aldrich [38] described these two kinds of allelopathy as true and functional allelopathy. Allelochemicals that released into soil can (1) effect directly on organisms, (2) can be that degraded or transformed with effect of soil microorganisms, (3) a third species may be induce to produce another compound which interferes with donor plants and (4) cause changes to soil abiotic factors that affect target plants.

It is impossible to separate direct from indirect allelopathic effects in field conditions and to assert that direct allelopathy is solely responsible for an observed phenomenon in the field, because many abiotic and biotic soil factors influence the fate of allelochemicals. Therefore, indirect allelopathic interactions, from an ecological point of view, are probably more important in plant communities than direct ones [36].

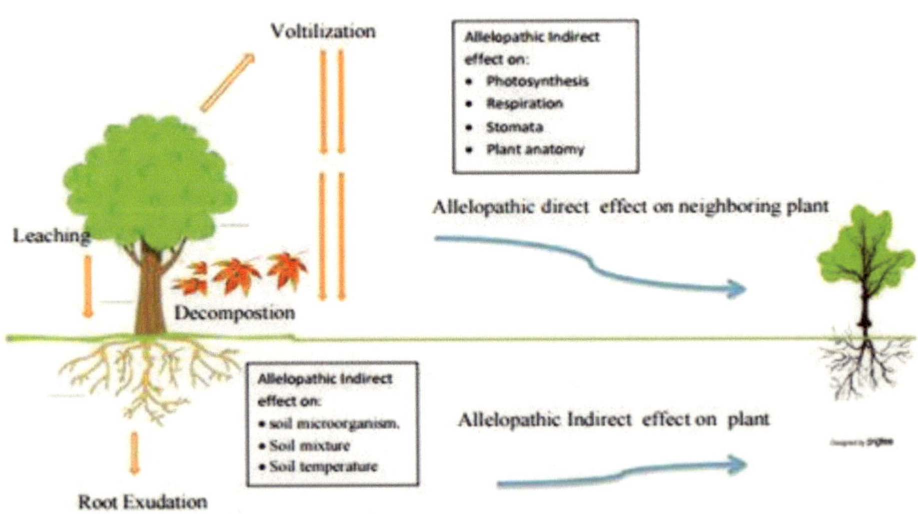

Figure 1.
Methods of released allelopathic compound in to environment.

4. Conclusions

Allelopathy is an important mechanism of plant interference mediated by the additional phytotoxins to the environment; chemicals with allelopathic potential are present in virtually all plants as in most tissues. Under appropriate conditions, these chemicals may be released into the environment, in sufficient quantities to affect neighboring plants allelopathy can affect many aspects of plant ecology, including growth, plant succession, the structure of plant communities, dominance, diversity, and plant productivity.

Acknowledgements

The authors is very grateful to the University of Mosul/College of Science for their provided facilities, which helped to improve the quality of this work.

Author details

Wasan S. Hussain[1*] and Mahmoud M. Abbas[2]

1 Biology Department, Collage of Science Mosul University, Iraq

2 Plant Nutrition Department, National Research Centre, Cairo, Egypt

*Address all correspondence to: wassbio54@uomosul.edu.iq

References

[1] Rice EL. Allelopathy. 2nd ed. Vol. 422. New York: Academic Press; 1984. ISBN 978-0-12-587058-0

[2] Hussain WS. The effect of malva residue in the germination and growth of eggplant and green pepper. Journal of the Faculty of Basic Education Researches. 2011;**11**(2):629-639

[3] Oudhia P, Kolhe SSS, Tirpathi RS. Allelopathic effect of *Blumea lacera* L. on rice and common kharif weeds. Oryza. 1998;**35**:175-177

[4] Cheema ZA, Farooq M, Khaliq A. Application of allelopathy in crop production: Success story from Pakistan. In: Allelopathy: Current Trends and Future Applications. Pakistan; 2012. pp. 113-143

[5] Farooq M, Bajwa AA, Cheema SA, Cheema ZA. Application of allelopathy in crop production. International Journal of Agriculture and Biology. 2013;**15**:1367-1378

[6] Aslam F, Khaliq A, Matloob A, Tanveer A, Hussain S, Zahir ZA. Allelopathy in agro-ecosystems: A critical review of wheat allelopathy-concepts and implications. Chemoecology. 2016;**27**:1-24

[7] Batish DR, Lavanya K, Singh HP, Kohli PK. Phenolic allelochemicals released by Chenopodium murale affect growth, nodulation and macromolecule content in ckickpea and pea. Plant Growth Regul. 2007;51:119-128

[8] Tanveer A, Rehman A, Javaid MM, Abbas RN, Sibtain M, Ahmad AU, et al. Allelopathic potential of *Euphorbia helioscopia* L. against wheat (*Triticum aestivum* L.), chickpea (*Cicer arietinum* L.) and lentil (*Lens culinarisMedic.*). Turkish Journal of Agriculture and Forestry. 2010;**34**:75-81

[9] Abbas MM, Hussain WS. Allelopathic Potential of Aromatic Plants on Germination and Growth of Some Ornamental Plants. Germany: LAP Lambert Academic Publishing; 2020. 58. p. ISBN 978-620-3-02499-9

[10] Al-Jihashi WS. The Biological Activities of Allelochemicals of Sunflower *Helianthus annuus* L. in Different Growth Stages. Iraq: Mosul University; 2005

[11] Malézieux E, Crozat Y, Dupraz C, Laurans M, Makowski D, Ozier-Lafontaine H, et al. Mixing plant species in cropping systems: Concepts, tools and models: A review. Sustainable Agriculture. 2009;**29**(1):329-353

[12] Macias FA, Oliva RM, Varela RM, Torres A, Molinillo MG. Allelochemicals from sunflower leaves cv. Paredovick. Phytochemistry. 1999;**52**:613-621

[13] Chou CH. Roles of allelopathy in plant biodiversity and sustainable agriculture. Critical Reviews in Plant Sciences. 2010;**18**(5):609-636

[14] Steudel B, Hector A, Friedl T, Lofke C, Lorenz M, Wesche M, et al. Biodiversity effects on ecosystem functioning change along environmental stress gradients. Ecology Letters. 2012;**15**:1397-1405

[15] Tilman D, Reich PB, Isbell F. Biodiversity impacts ecosystem productivity as much as resources, disturbance or herbivory. Proceedings of the National Academy of Sciences. 2012;**109**(26):10394-10397

[16] Lithourgidis AS, Damalas CA, Dordas CA, Vlachostergios DN. Annual intercrops: An alternative pathway for sustainable agriculture. Australian Journal of Crop Science. 2011;**5**(4): 396-410

[17] Callaway RM. Positive Interactions among Plants. Botanical Review. 1995; **61**(4):306-349

[18] Zhang F, Li L. Using competitive and facilitative interactions in intercropping systems enhances crop productivity and nutrient-use efficiency. Plant and Soil. 2003;**248**:305-312

[19] Chapagain T, Riseman A. Barley–pea intercropping: Effects on land productivity, carbon and nitrogen transformations. Journal of Field Crops Research. 2014;**166**:18-25

[20] Wang Q, Ruan X, Li ZH, Pan CD. Autotoxicity of plants and research of coniferous forest autotoxicity. Scientia Silvae Sinicae. 2006;**43**:134

[21] He Y, Ding N, Shi JC, Wu M, Liao H, Xu JM. Profiling of microbial PLFAs: Implications for interspecific interactions due to intercropping which increase phosphorus uptake in phosphorus limited acidic soils. Soil Biology and Biochemistry. 2013;**57**:625-634. DOI: 10.1016/j.soilbio.2012.07.027

[22] Alrababah MA, Tadros MJ, Samarah NH, Ghosheh H. Allelopathic effects of *Pinus halepensis* and *Quercus coccifera* on the germination of Mediterranean crop seeds. New Forests. 2009;**38**(3):261-272

[23] Shah AN, Iqbal J, Ullah A, Wu Y. Allelopathic potential of oil seed crops in production of crops. A review. Environmental Science and Pollution Research. 2016;**23**(15):116

[24] Caamal-Maldonado JA, Jimenez-Osornio JJ, Torres-Barragan A, Anaya AL. The use of allelopathic legume cover and mulch species for weed control in cropping systems. Agronomy Journal. 2001;**93**:27-36

[25] Hussain WS. Effect of spraying aqueous extracts of some crops plants on growth of four types. Plant Archives. 2020;**20**(1):1460-1464

[26] AL-Jehaishy WSH, The use of plant waste in the biological control of some bushes and their allelopathic effects on growth, and some physiological and anatomical traits [PhD thesis]. In Arabic: College of Science/University of Mosul; 2017

[27] Mandel RC. Weeds, Weedicide and Weed Control Principles and Practices. Bikaner, India: Agro. Botanical Publisher; 2000

[28] Farooq M, Jabran K, Cheema ZA, Wahid A, Siddique KHM. The role of allelopathy in agricultural pest management. Pest Management Science. 2011;**67**:493-506

[29] Mallik MAB, Williams RD. Allelopathic growth stimulation of plants and microorganisms. Allelopathy Journal. 2005;**16**(2):175-198

[30] Abbas MM. Enhancement of the nutrients efficiency and productivity of tomato (*Lycopersicum esculentum Mill.*) plants by using nano silver. Plant Archives. 2020;**20**(2):4242-4244

[31] Mallik MAB, Williams RD. Allelopathic principles for sustainable agriculture. Allelopathy Journal. 2009;**24**(1):1-34

[32] Einhellig FA, Rasmussen JA. Effects of three phenolic acids on chlorophyll content and growth of soybean and grain sorghum seedlings. Journal of Chemical Ecology. 1979;**5**:815-824

[33] Narwal SS. Allelopathy in Crop Production. Jodhpur, India: Scientific Publishers; 1994

[34] Harms CL. Oplinger ES. Plant growth regulators: Their use in crop production. North Centrel Region Extension Publication, NCR303. U.S Department of Agriculture Cooperative

State Research Service. Available online. 1993

[35] Scavo A, Restuccia A, Mauromicale G. Allelopathy: Principles and basic aspects for agroecosystem control. In: Gaba S, Smith B, Lichtfouse E, editors. Sustainable Agriculture Reviews. Vol. 28. Cham: Springer; 2018. pp. 47-101

[36] Inderjit, Weiner J. Plant allelochemical interference or soil chemical ecology? Perspectives in Plant Ecology, Evolution and Systematics. 2001;**4**:3-12

[37] Kobayashi K. Factors affecting phytotoxic activity of allelochemicals in soil. Weed Biology and Management. 2004;**4**:1-7

[38] Aldrich RJ. Weed-Crop Ecology: Principles in Weed Management. North Scituate, Massachusetts, USA: Breton Publishers; 1984

Chapter 7

The Role of the Livestock Farming Industry in Supporting the Global Agricultural Industry

Muhammad Irfan Said

Abstract

The livestock farming industry is a strategic industry and has a very important potential for the advancement of the agricultural industry. The livestock farming industry is an industry that plays a role in providing protein food for most of the world's population. Not only as a food provider, but also having a very large contribution to the progress of agricultural cultivation in the world. The livestock industry contributes part of the needs of the agricultural industry through the provision of environmentally friendly organic fertilizers. The integration of the livestock farming industry and the agricultural industry is a mutually supportive system. The integration of plants with livestock is basically an agricultural system characterized by a close relationship between plant and livestock components in a farm in a certain area. The bio-mass waste products from the agricultural industry can be used as a source of animal feed or as raw material for compost. The combination of the livestock industry with the agricultural industry provides enormous economic value for the development of the livestock and agricultural sectors. In Indonesia, the concept of integration between the livestock industry and the agricultural industry has been proven to improve the standard of living and economic status of the community. Some important materials will be discussed in more depth in this paper, including: (1) integration between livestock and rice, (2) integration between goats and cacao plants, (3) integration of livestock with oil palm, (4) integration of ruminants with cassava plants, and (5) integration of cattle with horticultural crops. Utilization of livestock and agricultural waste can reduce environmental burdens. Livestock farming industrial waste in the form of manure can be combined with agricultural wastes. This waste can be used as raw material for organic fertilizers which can be used as a fertilizer provider to support the cultivation of food crops, horticulture and plantations. The development of livestock in an area can encourage farmers to fill their vacant land to be planted with forage. Guidance efforts that will be carried out will be oriented towards land conservation. This is done through the arrangement of forage planting. Therefore, this will have a positive impact on increasing forage production. This effort will ultimately support the improvement of feeding patterns, proper land arrangement and being able to reduce the rate of erosion. The existence of livestock will add a source of manure. This is certainly synonymous with increasing land fertility which can provide business opportunities in optimizing land use diversification. The role of livestock in the livestock industry is the main support and complement in the system of integration in the livestock industry-agricultural industry.

Keywords: Industry, livestock, farming, agriculture, integrated

1. Introduction

Indonesia, is one of the countries in Asia which has a large proportion of the population making a living in the agricultural sector. This is due to natural conditions, which mostly consist of wide expanses of land, considerable biodiversity, and a tropical climate. With this tropical climate, the sun's rays can continue to shine throughout the year.

Based on 2019 data, the total population in Indonesia reaches 268,074,600 people [1]. The number of farmers in Indonesia reaches 33,400,000 people or around 12.45% of the total population. Based on these data, as many as 8% are young farmers aged 20–39 years, and the rest are farmers aged >39 years. In line with the increasing population, of course the need for animal food consumption (meat) will also increase. The increasing demand for meat for the community will be directly proportional to the rate of demand for meat. Unfortunately, the rate of demand has not been supported by efforts to increase domestic beef production. Currently, the availability of national beef in the country is still experiencing shortages. Therefore, additional imported meat is needed. Based on the data, approximately 35% of the total national demand for beef has been met from the imported component. The availability of meat in the country is quite safe even though it has not been able to meet all the needs of the Indonesian population. In Indonesia, the largest livestock providers for meat are large livestock (cattle, buffalo, horses); small livestock groups (goats, sheep and pigs); poultry (broilers) and various livestock groups (rabbits, quails and pigeons). In 2019, the livestock population for producing meat from large livestock groups amounted to 16,930,000 heads (cattle); 1,134,000 head (buffalo) and 375,000 head (horse). Meanwhile, the population of small livestock groups is 18,463,000 heads (goats); 17,834,000 head (sheep); 8521 head (pigs). Potential sources of meat obtained from the group of poultry (broilers) amounted to: 3,169,805,000 heads. In addition to the potential for producing meat, the potential of livestock as a producer of milk also has a very important role. This potential is obtained from the dairy farming business. The population of dairy cows in Indonesia reaches 565,000 heads. In addition, the potential for egg-producing livestock is obtained from poultry with the respective population for native chickens (301,761,000 heads); laying hens (263,918,000 head); ducks (47,783,000) and muscovy ducks (9446) [2].

The livestock sub-sector is one of the most important and strategic sub-sectors. This role is particularly in the process of meeting the needs of food sources of protein for the community. The demand for food from livestock in Indonesia continues to increase. On average, the consumption of animal protein in the Indonesian population is still low (<4 g/capita/day). The income elasticity of the demand for livestock products is relatively high. Meanwhile, the fulfillment of the need for beef is lower than the amount needed. This condition is an opportunity as well as a challenge for prospective breeders and livestock industry entrepreneurs.

In an effort to meet the needs of animal protein for the entire population, the role of the livestock sub-sector as a provider is very urgent. The need for animal protein is obtained from meat, milk and eggs. The availability of domestic meat has not been able to be fulfilled as a whole. Therefore, efforts to import meat are still an alternative. For milk and egg products, the needs of the people in the country are still relatively sufficient.

In 2020, meat production in Indonesia will reach 425,978 tons. In 2021, it is estimated that the demand for meat will increase significantly to reach 696,956 tons. Apart from domestic production, there is still a remaining supply of imported beef and buffalo as well as live cattle which are equivalent to meat. The amount of meat reached 47,836 tons, so that the total meat stock reached 473,814 tons [3].

Improvements in livestock governance need to be carefully planned. Fulfilling the needs of livestock and improving management is expected to increase livestock productivity. This effort can be made through the application of the concept of integration between the livestock industry and the agricultural industry.

The concept of an integrated farming system (IFS) is one that has been implemented since ancient times. This concept is characterized by the close relationship between the agricultural industry and the livestock industry. In principle, this system will ensure the realization of a sustainable agricultural system. Based on the history of agriculture, the integrated farming system has been abandoned since the end of World War II. By this time, specialized farming systems and monocultures had developed rapidly. This system has received full support, especially in relation to the uncontrolled use of chemical fertilizers, the use of herbicides to control weeds, as well as the use of pesticides and fungicides in controlling pests and plant diseases. In addition, the application of modern agricultural tools has also replaced traditional farming systems.

In general, these changes in agricultural practice patterns have indeed succeeded in significantly increasing yields and productivity of agricultural businesses. However, in the long term, this practice does not show a sustainable effort and ends with a decrease in land quality, an increase in the process of land and water pollution, the emergence of certain variants of pests and plant diseases.

As an agricultural country, the population in Indonesia has utilized most of the natural resources of its life to meet their daily needs. Agricultural products are a basic need and fundamental for the continuity of human life. Agricultural products referred to are products from the activities of the agricultural industry itself, plantations, livestock and fisheries.

2. Integration of cattle with rice plants

Replace Rice (*Oryza sativa* L) is one of the most important cultivated plants in the history of human civilization. Rice is thought to have originated from India or Indochina and entered Indonesia by ancestors who migrated from mainland Asia around 1500 BC [4].

Several countries in Asia are rice producers such as China (28% of total world production). Then followed by India (21%), and Indonesia (9%). However, only a small proportion of world rice production is traded between countries (only 5% -6% of total world production). Thailand is a major rice exporting country (26% of the world's total traded rice). Next followed by Vietnam (15%) and the United States (11%). Indonesia is one of the rice importing countries (14% of the world's traded rice) as well as Bangladesh (4%) and Brazil (3%) [5].

Rice plants are one of the staple foods for most of the population in several countries in Asia, including Indonesia. The use of rice as a staple food has been going on for hundreds of years. Currently, almost 90% of rice production has been produced and consumed by the occupation countries in Asia. This is certainly a sign that the efforts made by humans to increase rice production are uncertain. One of the reasons is that the world's population will continue to grow at a rate of 1.3% per year. It is estimated that in 2025 the world's population is estimated to reach 8.3 billion [6].

Based on data, in Indonesia, the population in 2010 has reached 237.56 million people. To meet these needs, the need for rice has reached 33.06 million tons/year, assuming 139 kg/capita/year. Related to the increase in the population which reached 241 million in 2011, the government has targeted rice production to be 68.59 million tons of milled dry rice (MDR) or the equivalent of 38.57 million tons of rice. This figure increased by 2.1 million tons of MDR or 3.2% compared to the target of rice production in 2010 [7].

Apart from rice, the commodity of meat is quite important. In the last 10 years, the development of the livestock sub-sector has shown tangible results, especially its contribution to gross domestic product (GDP). Total consumption of meat, eggs, and milk each increased by 7.6%; 5.22%; and 0.92%. However, the increase in consumption has not been matched by an increase in production, especially beef, whose population has even decreased to 4.1%/year [8]. To meet domestic demand for meat, part of this need must be imported. The demand for beef in 2014 refers to the consumption of meat per capita which has increased from 2013 of 2.2 kg/year to 2.36 kg/year. Meanwhile, the fulfillment of beef consumption originating from imported products is only 58,280 tons or 9.8 percent of feeder cattle amounting to 34,970 tons or equivalent to 175,407 head, while in the form of meat is 23,100 tons. To provide local meat needs in 2014, a cattle population of 19.7 million heads is needed, while for beef cattle, 17.6 million must be available, higher than the beef cattle population in 2013 of 16.8 million tons [9].

The beef self-sufficiency program has been launched since 2005 and is targeted to be achieved in 2010, but in reality it has not been achieved. Therefore, the government is re-targeting a new self-sufficiency and finally achieved in 2014. The concept of beef self-sufficiency is certainly not impossible to achieve if there is seriousness from all parties to develop domestic beef cattle agribusiness. Market potential and supporting resources should be an opportunity for the development of beef cattle with a comparative and competitive advantage in both local and export markets [10].

The cattle population in Indonesia can grow because the business is quite prospective. This is evident from the growing business of cattle fattening in a number of regions and the Organic Fertilizer Processing Unit (OFPU) Program since 2011. The farmer groups are given a number of cows and then the farmers process cow dung into compost. There are farmer groups that are able to develop cow dung and urine production as the main business, while cattle fattening is a side business. The sale of cow dung and urine provides an income of IDR 22,000/head/day, while the purchase of feed is only IDR 7000/head/day. The average compost price at OFPU ranges from IDR 600–750/kg [11].

The rice-livestock integration system (RLIS) has become part of the farming culture in Indonesia. This system is able to utilize local resources, namely by-products in the form of straw and bran, and livestock manure efficiently. The main characteristic of RLIS is that there is a link between plants and livestock, for example rice crop waste (straw) is used as animal feed, and vice versa, livestock manure can be used as organic fertilizer to meet plant nutrient requirements [12]. Therefore, the RLIS assessment program was initiated in conjunction with the rice paddy integrated crop management (ICM) assessment program. The ICM recommends the use of organic materials as one of the main components. Organic fertilizers are needed to increase rice yields, improve soil physical and chemical properties [13], and suppress the use of inorganic fertilizers [14]. Excessive use of nitrogen and continuous application of P fertilizer in some rice fields can damage the nutrient balance in the soil. The reduced content of organic matter in agricultural land in Indonesia today shows the need for efforts to increase the quality of soil organic matter content by two times to restore normal soil health conditions [15]. Nurawan et al. [16] stated that organic fertilizers can increase rice yield by 0.9 ton/ha compared to without organic fertilizers.

The RLIS provides benefits to farmers, namely cow manure and agricultural by-products in the form of straw and bran. Utilization of waste from livestock manure has not been optimal. It can be used as organic fertilizer to increase soil fertility or can be sold as a source of additional income. The available agricultural waste can be used as a source of quality feed so as to reduce the cost of providing feed [17]. The RLIS pattern is a solution to feeding problems. This pattern can strengthen food security through processing agricultural waste into high nutritional value feed and can meet the needs of cattle. In general, RLIS is one solution to achieve food security, independence and sovereignty.

The application of RLIS in several regions in Indonesia has had a significant effect on increasing farmers' income. The application of an integrated farming system encourages an increase in farmers' income. An overview of the benefits obtained through the implementation of an integrated farming system in Indonesia at 3 different provincial locations is presented in **Table 1**.

Based on the data in **Table 1**, it can be seen that the benefits of rice and livestock farming which are managed in an integrated manner will be able to provide higher profits than only in partial form. Farming that is done partially means farming that is only done singly (rice plants), while integrated farming means farming that combines rice plants with livestock. The straw waste generated from the rice cultivation business is then fermented and given to livestock. In Central Java Province, farming which is carried out in an integrated manner is able to provide a profit of 15.86% higher than that which is carried out partially. The same thing also happened to two other provinces, Bali Province and West Nusa Tenggara (NTB) Province at 29.19% and 27.72%, respectively. Another advantage can be seen in the component of the benefit cost ratio (BCR), which on average is higher in regions that implement

	Partial			Integrated		
Description	**Rice**	**Cattle**	**Total**	**Rice**	**Cattle**	**Total**
1. Jawa Tengah Province						
a. Revenue	6368	4232	10,600	6754	4527	11,281
b. Profit	2958	397	3355	3324	574	3887
				(12.37)	(44.58)	(15.86)
c. Benefit Cost Ratio (BCR)	1.87	1.12	1.47	1.96	1.17	1.57
2. Bali Province						
a. Revenue	6666	5051	11,717	7246	5315	11,281
b. Profit	3283	146	3429	4019	411	4430
				(22.42)	(181.5)	(29.19)
c. Benefit Cost Ratio (BCR)	1.97	1.07	1.45	2.25	1.22	1.86
3. Nusa Tenggara Barat (NTB) Province						
a. Revenue	5203	4497	9700	5360	4723	10,083
b. Profit	1843	329	2172	2207	567	2774
				(19.75)	(72.34)	(27.72)
c. Benefit Cost Ratio (BCR)	1.55	1.16	1.34	1.70	1.30	1.50

Table 1.
Comparison of Revenue and Profits from Farming through the Application of the Rice–Livestock Integration System (RLIS) in 3 (three) Provinces in Indonesia [18].

an integrated system respectively (1.57; 1.86 and 1.50) than partially (1.47; 1.45 and 1.34) for the three study areas. A higher BCR value indicates that an area that implements an integrated system has a high effectiveness in the use of inputs or production costs. Based on the aspect of market demand, there is a tendency for consumers to be more dominant in choosing organic agricultural products (using organic fertilizers and free of pesticides). This is probably due to considerations of health factors, although at a relatively higher price.

Based on the data in **Table 1**, the first assumption used is the cost of fertilizers. For farming that is managed in an integrated manner, the cost of fertilizer used is only around IDR 500,000-600,000/Ha. On the other hand, partially farming requires fertilizer costs ranging from IDR 621,000-733,000/ha. The use of organic fertilizers from livestock waste in integrated farming can save fertilizer costs by around 18.14–19.48% or around 8.8 percent of the total cost. The second assumption is the cost of feed. An integrated livestock business, rice crop waste in the form of straw can be used as feed. The cost required is only IDR 410,000-889,000/head, while partial farming (without hay feed) costs IDR. 735,000-1,377,000/head. Therefore, this cost can be assumed that farming businesses that utilize agricultural waste can save labor costs by 35.44–44.22% or around 5.26–6.38 percent of the total cost of livestock farming.

Since independence, Indonesia has not been able to meet food needs from its own production. Therefore, importing several foodstuffs must be an option. The food sovereignty program is one solution to reduce the rate of imports of food products [19]. The food sovereignty program is able to reverse the concept of modernization into things that are back to basic or back to nature by prioritizing production for the fulfillment and sustainability of local food and local markets through the provision of agricultural production inputs that utilize local wisdom and are environmentally friendly [20]. One thing that needs to be known is that the main principles and problems that arise in the food production process are self-reliance and self-sufficiency. As an effort to achieve self-sufficiency in food production, of course, an economic democracy is needed. For the government, they must take redistribution actions. Special parties such as the government must ensure the people's livelihoods and income by providing capital equitably in the agricultural sector. Matters related to the food production program, must rely on efforts to mobilize the largest portion of small-scale food producers. It is focused on marginal food producing sectors, providing access to resources such as land, water, seeds and livestock [21].

It is known that, Indonesia has absolute advantages (comparative and competitive) in building food independence and sovereignty. Of course, this potential has been characterized by several things, such as: (1) Indonesia is characterized by being on the equator with a tropical climate and a very conducive sunlight intensity for agricultural production; (2) availability of land and water so that it is not a limiting factor which is quite worrying; (3) availability of living natural resources accompanied by local food diversity; (4) advances in food production technology, including those that have developed in Asia; and (5) the government's 15 million ha perennial agricultural land program has not been implemented properly [17].

Food sovereignty and food self-sufficiency programs always require a strong food security. This includes food availability, accessibility, price stability, utilization, quality, and safety [22]. As an effort to create food sovereignty, the government needs to make efforts to protect its farmers. In addition, the government always encourages farmers to be ready to compete with food corporations. Of course, this will have a negative impact on farmers. A state subsidies program by protecting trade and providing adequate subsidies to farmers must be implemented.

The government must allocate an adequate subsidy budget. The percentage of the food subsidy budget must be greater than gross domestic products (GDP) [23].

One alternative strategy that is quite prospective in building food independence and sovereignty is to utilize local resources that are supported by agriculture-based industries, soft credit schemes, and infrastructure development in rural areas. This is predicted to be able to create a market for primary agricultural production and new jobs in rural areas [17].

The concept of integration of the livestock industry with the agricultural industry, especially rice, can be seen from their respective roles. The role of livestock (beef cattle) in the rice cultivation process acts as a land processor by generating energy. Livestock waste in the form of manure released during the soil processing process will be decomposed so that it can fertilize the hardened soil due to the use of very intensive chemical fertilizers. In his calculations, an adult cow can produce approximately 8–10 kg of manure per day which can be processed into 4–5 kg of compost/day [24].

Liquid waste from livestock in the form of urine is collected from livestock disposal. Then stored in a plastic drum, processed with additional ingredients which are then deposited. Liquid fertilizer from urine can be used to fertilize rice plants through leaf spraying techniques. Organic fertilizers from feces and urine are much desired by consumers, but the amount is still very limited [25].

The existence and role of livestock waste has not completely replaced the position of chemical fertilizers, however, the deficiency of several nutrient compounds in chemical fertilizers can at least be complemented by organic fertilizers from livestock manure [26]. The cattle culture management system which is carried out intensively by considering the aspects of feed (concentrate and fermented rice straw), collective cage management, and animal health can increase the average daily gain (ADG) of 0.89 kg/head/day during the fattening period. This value is higher than the farmer method which is only 0.29 kg/head/day. The resulting ADG increased by about 0.6 kg/head/day (67.42%), so that it was able to produce ADG from 0.29 to 0.89 kg/day. The process of fattening cattle is not only for achieving high ADG values, however, how can cattle use rice straw as agricultural waste which has not been optimally used. This of course will be able to reduce farmers' expenses from production costs and most importantly environmentally friendly [27].

The level of livestock productivity is influenced by at least 70% from environmental factors, and the remaining 30% is genetic factors. Feed (up to 60%) is one of the environmental factors that play the biggest role in productivity. Based on this, it can be said that, although livestock have prime genetic potential, but not supported by quality and available feed, livestock productivity will be difficult to achieve. The important role and contribution of the rice plant business in the management of cattle feed is the presence of agricultural waste in the form of rice straw (whether or not fermented). However, rice straw that does not undergo a fermentation process, of course, cannot be fully digested by livestock. This is caused by the high lignin and hemicellulose compounds in the straw waste [28].

The biodegradation process of lignin compounds in straw aims to remove lignin, increase the digestibility of cellulose so that the quality of straw as animal feed will increase. The use of TLiD and BOpR bacterial isolates is able to degrade lignin and organochlorin (lignolytic). These isolates are specific for growth on rice straw. Application of TLiD and BOpR isolates in the fermentation process of rice straw can reduce the lignin content of rice straw up to 100% on the 7th day of fermentation and increase the crude protein of rice straw. High degradation efficiency of isolates where lignin degradation is higher than cellulose [29]. Processed agricultural waste (by-product) has a protein content of 12% higher than grass protein content of around 9%. The palatability of processed feed is better because it contains molasses and pikuten (commercial minerals) [30].

Rice plant waste in the form of straw needs to be utilized optimally. The straw burning activity will reduce the nutrients contained in it. The straw that is stored in the land without fermentation is very difficult to be bound by soil particles. Therefore, this waste requires special attention so that its quality can be improved. Compost processing from straw can be done in two ways, namely: 1) stacked and turned and 2) stacked with ventilation without turning. The decomposition process can be accelerated by using a decomposer in the form of microorganisms. Several commercial decomposers are known to contain several kinds of microbes, for example M-Dec products containing *Trichoderma harzianum*, *Aspergillus sp*., and *Trametes sp*. Orgadec products contain *Trichoderma pseudokoningi*, and *Cytophaga sp*. EM-4 products contain photosynthetic bacteria, lactic acid, actinomycetes, yeast, and fermented fungi [31].

Probion product is a type of animal feed additive that can be used directly as a concentrate feed mixture to improve the quality of rice straw through the fermentation process. Probion's product is a consortium of microbes from ruminants that have been enriched with essential minerals for microbial growth needs. Ripe compost is characterized by a temperature that is already constant (40–50°C), crumbs, and has a dark brown color. The compost obtained is ± 500 kg with C-organic quality >12%, C/N ratio 15–25%, 40–50% moisture content, and light brown-black color [32].

Rice straw as agricultural waste has abundant potential, but has not been utilized optimally as cattle feed. One of the obstacles is the high crude fiber content, but very low protein content and digestibility. The use of straw directly or as single feed for livestock certainly cannot meet the nutritional value needed by livestock during the production process [33]. Utilization of straw waste from rice plants as animal feed is one of the efforts in realizing an integrated system of rice plants with livestock. A complete description of the integration model was presented in **Figure 1**.

The nutritional content of straw can be increased through the fermentation process using probiotics as a bio-degradation. The fermented straw has a protein content that almost matches the quality of elephant grass. The fermented straw should be stored as soon as possible in a dry place so that its quality and quality are maintained. The potential of straw as animal feed is able to streamline the labor

Figure 1.
Utilization of straw waste as animal feed to support the integration of rice plants with cattle. Source: https://banjarmasin.tribunnews.com/2015/09/02/peternak-ganti-rumput-dengan-jerami.

that farmers have to prepare to find grass. In fact, the results of the study show that providing agricultural straw waste to livestock by adding microbes and urea has been shown to improve livestock productivity [34]. Rice plants will produce straw as a by-product. In each hectare, rice fields will produce fresh straw waste of 12–15 tonnes/ha/season. Furthermore, waste that has gone through the fermentation process will produce 5–8 tons/ha which can be used to meet the feed needs of 2–3 cows/year [14].

The use of microbes in the straw fermentation process is very effective in improving the quality of the straw. The nutritional composition of rice straw that has been fermented using a Starbio starter as much as 0.06% of the weight of rice straw generally shows an increase in quality compared to unfermented rice straw. The fermentation process is able to increase the crude protein content of rice straw from 4.23% to 8.14% which is then followed by a decrease in crude fiber content. These results indicate that, Starbio starter is a proteolytic microbe that can produce enzymes to break down proteins into polypeptides which then become simple peptides. The application of microbial starter was able to reduce the rice straw cell wall content from 73.41% to 66.14%. The lignocellulose and hemicellulose bonds of rice straw will be released during the fermentation process. The activity of lignolytic microbes in the microbial starter helps to break down the lignocellulose bonds so that cellulose and lignin can be released from these bonds by the activity of the lignase enzyme. The decrease in cellulose and lignin content during fermentation is one proof that these bacteria are working well. Lignin compounds in straw are a physical barrier that can inhibit the digestibility of enzymes in plant tissue. In addition, lignin binds closely to hemicellulose. The decrease in cell wall content indicates that the process of breaking down cellulose cell walls has occurred so that the feed will be more easily digested by livestock [35]. The fermentation results can increase the nutritional level of the straw Thus, the increase in live weight of livestock can increase very significantly. Cattle that are given additional feed such as straw and probiotics are able to give a live weight gain of 0.56–0.68 kg/head/day higher than the control [36].

One type of waste produced from the rice processing is bran. This bran consists of a layer of aleurone and a small portion of endosperm, pericarp, pigment, and germ [37]. The amount of bran that can be produced is as much as 8–10% of the weight of milled rice, so that its availability is quite abundant. Rice bran contains dry ingredients as much as 88.30%; crude fiber 15.30%; ash 9.90%, 10.10% crude protein, 4.90% crude fat, and 48.10% BETN [36]. Bran is a source of carbohydrates that are easily available and very effective in improving the quality of fermentation in rice straw [38]. Giving rice bran and Bioplas probiotics to pregnant cows of local cattle can increase the body weight of the cows by about 0.5 kg/head/day and can increase the birth weight of the calves by about 10.5 kg compared to the control 8.9 kg. Feed consumption has increased by about 5.2 kg. In addition, giving bioplas bran and probiotics to cows could re-estrus after 62 days after giving birth compared to controls about 85 days after giving birth [39].

Utilization of solid waste (manure) and liquid (urine) in cattle cultivation as organic fertilizer is expected to be used as a source of additional income for breeders. In addition, it can improve the fertility of agricultural land. The introduction of technology for the integration of livestock with rice plants was able to increase farmers' income by IDR 34,488,800, – higher than traditional technology of IDR 22,903,200, – Based on the results of the R/C ratio analysis, the value was 6, higher than the traditional pattern with R/C ratio of 4, so it is feasible to be cultivated by farmers. Rice farming which is integrated with cattle is an efficient and effective farming for improving the farming income of the people by selecting narrow land in rural areas. Farming with rice-cattle integration pattern can increase farmers' income by 70% on rice scale farming with 5 ha of plant area and cattle ownership of 20 heads [25].

The development of cattle using the RLIS method in several potential areas has a positive impact on increasing the domestic cattle population. The result is expected to be able to be self-sufficient in meat. This program aims to maintain the balance of local livestock stocks as highly valuable germplasm. In addition, it is also aimed at reducing the need for meat imports which have been very difficult to stem due to the high domestic demand for meat [40]. Another positive impact that occurs is the ability to improve the performance of other farmer groups in terms of buying and selling cattle. With a pattern of planting rice 3 times/year and being a technical irrigation area, the straw is entirely for animal feed needs. The provision of straw for animal feed is quite high, namely 25 kg/day/head for seed cows and 31 kg/day/head for fattened cows. Several obstacles faced in implementing the RLIS pattern, namely: (1) the working mechanism of the group did not work well; (2) the utilization of the collective pen is not optimal at all so that the level of utilization of the pen facilities is still low, (3) the process of mentoring and coaching is not effective because the position of the cattle cannot be kept in one collective pen. As a result, the existing livestock will be scattered according to the domicile position of the breeders. Another obstacle faced is that the transportation process from the compost location to the fields is still burdensome. On the other hand, the use of manure by group farmers is not yet entrenched. This requires socialization so that available manure can be utilized [41].

The RLIS program was initiated in conjunction with the ICM program. In addition, the development of the RLIS needs to be carried out through a farmer group approach to facilitate agricultural extension, adoption of rice-livestock technology, and government assistance channels. The advantage of the rice-livestock integration pattern is the use of potential plant waste as a source of animal feed, utilizing livestock manure as manure, creating new jobs in rural areas, and increasing community participation in realizing agribusiness that is competitive, environmentally friendly, and independent. Obstacles to SIPT in realizing food security include the working mechanism of farmer groups that has not been running well, the use of collective cages has not been optimal, the process of mentoring and coaching has not been effective because the location of livestock is scattered, the use of manure has not become a culture among rice farmers, and the implementation of RLIS is carried out throughout province and no clear progress. Future improvements in RLIS should be focused on areas of production centers so that they are large-scale and have a significant impact on population growth and livestock productivity. The processing of livestock waste is close to the rice planting location to minimize transportation costs so as to create zero waste and a good integrated farming system to realize food sovereignty [42].

Rice-livestock integration technology can improve additional income for farmers in a condition of synergy. Utilization of farm input from sources has been available optimally. Products produced during the rice production process have been integrated with cattle fattening, including straw and feed which have economic value. Cow manure is used through the recycling process into biogas. Worms and organic fertilizers are used for fertilizing plants, while livestock waste is used to fertilize fish ponds. Paddy fields, in addition to the main products of rice, also produce bran and straw which can be used as animal feed. In this case, all waste, both livestock and plants, has added value and does not pollute the environment.

The RLIS in the agricultural system is a very important strategy to realize environmentally friendly farming, the welfare of farmers and rural communities. RLIS is one of the government's programs to realize food sovereignty which has become the right of all Indonesian people to obtain food that is healthy, sufficient, and easily accessible for survival. The principle of RLIS is farming that applies the concept of zero waste by utilizing local resources, namely rice straw, bran, and livestock manure efficiently. The potential for agricultural by-product (by-product) in the form of straw is very abundant, but it has not been used as cattle feed for cattle. The nutritional value of straw can be increased through probiotic fermentation and almost matches the quality of elephant grass. The technology of processing livestock manure into compost is an alternative solution to environmental problems that can solve critical rice field problems. The integration of cattle and rice farming is an efficient and effective farming for improving the farming income of the people with narrow land ownership in rural areas.

3. Integration of goats with cocoa plants

The availability of agricultural land as the main capital for farming tends to decline from year to year. This is due to the rapid development of the population and changes in regional spatial patterns. As an impact, the production system changes. This will also have an impact on declining farming businesses.

Another problem that arises is the decrease in soil productivity. This is caused by soil erosion and nutrient leaching. The major impact that may occur is the increasing costs and dependence on external inputs (chemicals and energy). In addition, there is an increasing threat of agrochemical residues to food quality and safety due to the increasing activity of water pollution (fertilizers and pesticides).

The potential for the development of the estate sub-sector to support the development of modern livestock businesses as a source of feed is developed through various systems. The integration pattern of livestock and plantation products in the form of utilization of staple crop waste and intercropping waste is very interesting to be developed.

In the last 2 decades, the development of cocoa cultivation has developed rapidly. Plantation statistics show that in 1997, in Indonesia, the land area for plantations was 529,057 ha. However, in 2015 the land area had reached 1,722,315 with production reaching 760,429 tons. Indonesia ranks as the 3rd largest cocoa bean producing country after Pantai Gading and Ghana. These three countries have contributed more than 70% of the world's cocoa needs. In Indonesia, the cocoa plantation industry is spread across various provinces, such as South Sulawesi, Southeast Sulawesi, West Sulawesi and Central Sulawesi. Cocoa is one of the main plantation commodities in Indonesia, which ranks as the third largest source of non-oil and gas foreign exchange after rubber and palm oil commodities. Most of the cocoa farmers support their families from this commodity. However, there were significant problems. The productivity of the cocoa plant is still very low (650 kg/ha/year). Farmers sell their produce in the form of cocoa beans in the form of dry beans. In addition, farmers also cultivate cocoa only in monoculture form and do not make efforts to diversify products, so that farmers' income is very minimal. On the other hand, the production costs for cocoa cultivation are getting bigger, especially in meeting the needs for fertilizers and pesticides [43].

Livestock farming that has been widely developed by rural communities is goats. Indonesia is one of the countries with a tropical climate, has a climate type suitable for the development of goats, extensive land and forage production that is far from sufficient to raise 100 million goats or 10 times the current population of goats. On the other hand, domestic marketing of goats has reached a saturation point. The amount of supply of goat meat is greater than the amount of demand.

One of the problems faced by goat breeders is the availability of feed which is quite far from the location of the pen. So far, livestock feed has been obtained from the forest edge. There is no planting of animal feed. Forage crops are allowed to grow wild without any maintenance. The livestock feed is further away from the village or location of the goat pen. As a result, 1 farmer is only able to raise 2–3 goats on average.

As an effort to develop the population of ruminants (goats), it is no longer possible to rely solely on grass or grazing sources. Land for growing feed is increasingly limited. On the other hand, the potential for alternative feed sources for ruminants (goats) is very large, especially fiber feed sources. The potential for fiber feed can come from the by-product of the agricultural and plantation industries. The use of agricultural/plantation byproducts as feed ingredients is a wise action to create food security based on local resources and help reduce environmental pollution [44].

Based on these problems, of course we need a system that can support mutually beneficial synergy. It is hoped that this pattern and system will provide benefits to both parties. One possible synergy pattern that can be done is the integration pattern between goat livestock and the cocoa plantation industry. Steps that must be taken in anticipating a sustainable farming system are diversification (multi-commodity) farming. One of them is implementing the integrated farming model of plants and livestock. This is one of the alternatives in carrying out business efficiency on a relatively fixed land area, but it can increase business productivity. This activity will provide added value from various business sectors that support each other. Integrated agriculture (crop-livestock integration) is an agricultural system characterized by a close relationship between plant and livestock components in a farm or within an area. One of the characteristics of this relationship is the existence of various resources such as forages, plant residues, and organic fertilizers produced by livestock in a production process. The most important thing that needs to be understood from the concept of crop-livestock integration is its ability to stop agricultural practices that destroy land resources and reduce agricultural productivity. The impact of these activities is that farmers can slowly escape poverty.

Due to the large population with a growth rate of around 1.5% and the increasing elasticity of demand for livestock production, of course, the need for ruminant livestock production (meat, milk and skin) will also increase. In recent years, the number of imports of live livestock and frozen meat along with milk and skins has increased to meet domestic demand. Therefore, the level of economic disparity between domestic demand and production capacity will increase. The domestic market will certainly be the target of producing countries. This can happen if there are no serious efforts to increase domestic production.

The potential for ruminant livestock development can be integrated with various types of plants, fish and forests. Therefore, if this potential can be utilized, the shortage of domestic supply can be met. The excess production can then be exported. Most of the breeders are rice field farmers, garden workers, cultivators and fishermen. Therefore, the combination of livestock business activities with other farming activities can increase business efficiency. The competitiveness of the products will also experience a very significant increase [45].

The system of integration of crops and livestock in the Sahel, Africa has been carried out in everyday life by farmers [46]. Farmers have considered climate risks, economic factors and livestock health. Although the integration pattern has its drawbacks, however, they can anticipate through various strategies based on livestock mobilization as well as regulation or diversification of activities.

In Indonesia, the cocoa plantation industry is dominated by smallholder plantations. Cocoa farmers usually use the waste of cocoa pod husks and forage from protective plants (*gamal* and *lamtoro*) as feed for goat farming. Cocoa pod husk waste is always available because cocoa pods on smallholder plantations can be harvested almost all year round. Meanwhile, with the correct intervals and cutting methods, forage from *gamal* and *lamtoro* plants as cover crops in cocoa plantations is also a readily available feed ingredient. However, these activities have not been managed optimally and efficiently.

Goat farming with a cage system in the cocoa plantation area will produce solid material in the form of goat manure. This material comes from goat manure and can be used directly in plantation areas. This will spur an increase in production and productivity of cocoa plants in each hectare. In fact, this can certainly reduce the costs incurred by farmers to meet the needs of production facilities such as chemical fertilizers. In addition, goats that are grazed in the cocoa plantation area will utilize grass and weeds around the cocoa trees. This of course will greatly save the use of pesticide costs and the cost of maintaining the garden. Farmers do not need to look for food because the plantation area has grass and waste from cocoa plants such as cocoa shells. This waste can be used by goats as animal feed. Thus, the daily activities of farmers looking for grass can be diverted to other activities that are more useful. The model for implementing the integration of cocoa plants with goats is clearly presented in **Figure 2**.

Cocoa pod husk (CPH) has the potential as an alternative feed source for ruminants. The potential for CPH in Indonesia is quite large, both in terms of quantity and quality. The availability of CPH during the harvest season is very large and able to meet the needs of 635,305 livestock units per year. The husks as a source of fiber feed can replace grass. CPH contains crude protein 6.80–13.78%; neutral detergent fiber (NDF) 55.30–73.90% and acid detergent fiber (ADF) 38.31–58.98%. In addition, CPH also contains anti-nutritional compounds, including lignin, tannins and theobromine. In an effort to optimize the potential of CPH as feed, processing is needed to increase nutrient value and digestibility, reduce the content of anti-nutritional compounds so that the negative effects of antinutrient compounds can be minimized and extend shelf life. Through a system of integrating plants with livestock, environmentally friendly agriculture can be developed, agricultural systems can run sustainably, optimal use of resources, human health status will be improved and maintained, the environment will be protected and foodstuffs will be fulfilled. Goats get their feed from parts of the cocoa plant (pruned cocoa leaves,

Figure 2.
Utilization of cocoa waste as animal feed to support the integration of cocoa plants with goats. Source: https://dpkp.jogjaprov.go.id/baca/Pemanfaatan+Limbah+Kakao+Untuk+Pakan+Ternak+Mendukung+Integrasi+Kakao-Ternak/190221/e84ce4d623c0e94ef3b54c3d9ec9d7dd1313653ca54a3e8aef9b95204b6ce0e9280.

husks and leaves from pruned protective plants such as *Gliricidia* or Lamtoro. Based on their nutritional content, feed ingredients (*Gliricidia* or Lamtoro) are quality feed ingredients. Crude protein content of the CPH are around 10% while forages from Gamal and Lamtoro are more than 20% [47].

4. Integration of cattle with palm plants

The low productivity of cattle has forced the government to import, especially heifers. This of course raises concerns on the part of the government. Total imports of heifers reached 800,000 head and frozen meat equivalent to 781,117 head of cattle. The total national population of beef cattle in 2016 reached 15.4 million heads [48].

The beef cattle rearing system is generally still done conventionally. Sources of forage are obtained from grazing land. This feed is a mainstay of breeders as a source of forage, so that the productivity of cattle is low. The oil palm plantation industry is an industry that can be integrated with the livestock industry, especially related to beef cattle cultivation. In Indonesia, local beef cattle has a number of advantages. One of the advantages is the ability to consume high fiber feed such as forages and concentrates in large quantities [49]. An illustration of a model that integrates cattle with palm plantations was presented in **Figure 3**.

Types of feed with high fiber content are also available in the palm oil industry. Several by-products from the oil palm plantation industry can be used as cattle feed. The land area is directly proportional to the potential use of the oil palm plantation industry by-products. The wider land will increase the use of by-products from the plantation industry. in 1 ha of oil palm land will be able to accommodate 3–5 adult cows [50].

The amount of waste from the oil palm plantation industry in the form of fronds and leaves that can be obtained for every hectare of oil palm plantation reaches more than 2.3 tons of dry matter. This amount is obtained with the assumption that each oil palm tree can produce 22 fronds and fresh leaves/year. Weeds that grow in the form of grass around trees can also be used by cows as a source of forage [51] . In every hectare of industrial land for oil palm plantations, grass or weeds that grow around the plant can reach 3–5 tons/year. In the first 2 years of planting, legumes can reach 5–7 tons of dry matter/year. After that, the production will decrease 1–5 tons/year at the age of 2–5 years [52].

Figure 3.
A farming system model that integrates cattle with palm plants. Source: https://www.agrofarm.co.id/2019/10/19517/.

Several years ago, in several provinces in Indonesia, the government has initiated an integration concept between the livestock industry, especially cattle, and the oil palm plantation industry. This concept basically aims to increase meat production. In addition, to optimize natural resources in order to support one another. There are several efforts that can be made to optimize meat production including: 1) optimizing the use of critical land, 2) optimizing the use of alternative feed material sources and 3) integrating the livestock industry with the plantation industry, especially oil palm plants. Efforts to combine the two industries are then known as LEISA (Low External Input Sustainable Agriculture) [53]. This concept is a symbiotic mutualism between livestock and plantation crops which can provide benefits to both parties. The application of this concept includes: 1) utilizing waste from oil palm plantations in the form of solid waste, midrib and meal as a source of animal feed ingredients, 2) waste from livestock in the form of manure and waste from oil palm plants (non-feed) then formulated and processed into compost. The compost will later be used as organic fertilizer which can improve the physical, chemical and biological properties of the soil in the plantation area, and 3) weeds that grow wild among oil palm plants can be used by livestock (cows) as a source of feed so that it will improve the productivity of oil palm plants [54].

In the future, livestock business development activities are expected to be able to change the view of breeders from a production system to a livestock system with an optimally integrated agribusiness concept. The concept of agribusiness is a development concept in the livestock sector which is a system and consists of several sub-systems, namely; (1) upstream agribusiness sub-system, namely the entire economic activity that produces production facilities (breeding industry and feed industry); (2) on-farm agribusiness, namely all activities related to livestock cultivation; and (3) down-stream agribusiness, namely all economic activities related to the processing of primary livestock commodities into processed products (livestock processing and marketing industries); and (4) supporting system, namely all activities that provide services from the three agribusiness sub-systems [55].

According to [56], the application of an integrated business system between cattle and the oil palm plantation industry is expected to be part of a farming system that can be carried out in a sustainable manner. The pattern and system of integrating cattle with oil palm plantations are expected to reduce the problem of waste from cows (manure and urine) and waste from oil palm plantation business activities. Efforts to use land must be based on: (1) land as a source of animal feed; (2) all types of land are focused as a source of animal feed; (3) The use of land for livestock can be interpreted as an attempt to harmonize the land use with the agricultural business system. The potential for beef cattle development still needs to be improved through technological innovation and the implementation of the concept of integration of crop livestock with livestock (crop livestock system) through optimizing the use of agricultural waste for feed and the use of manure for organic fertilizer for agricultural crops.

5. Integration of ruminants with cassava plants

The livestock development program is one of the important aspects in the agricultural development program, so that the role of farmers and breeders will determine the success of this development. In an effort to meet these expectations, the main challenge facing the livestock sector today is how to produce livestock products that are highly competitive in terms of quality, quantity, variety and price. This can be done by optimizing the use of local feed so that domestic and global market needs can be met. In Indonesia, both the potential and available land for expansion of agricultural areas are still quite large. However, this requires caution because the need for land use for agriculture and non-agriculture is currently increasing. The impact that occurs is of course the existence of competition in land use.

Based on data, Indonesia is the second largest cassava producing country with a production of 24 million tons after Nigeria country (52.4 million tons) [55]. In 2011, cassava production in Indonesia has reached 24 million tons, which is produced from a harvest area of around 1.2 million hectares of land with a productivity of 20 tons/ha [57]. This production results in cassava being the second largest production after rice (65 million tons) [58]. The development of cassava growth over the last 10 years shows that the area of cassava plants is relatively constant and even tends to decline in 2012 (−0.56%/year), however, the production has actually increased with a growth rate of 2.62% /year. This is due to improved productivity and shows an increase with an average growth rate of 3.97%/year [59].

The cassava plant (*Manihot esculenta*/*Manihot utilisima*) is a root tuber plant that has an elongated shape with a center line of about 2–3 cm and a length of 50–80 cm, depending on the type and variety. The plant consists of several parts, namely tubers, stems, roots, flowers and leaves. The tuber part of cassava can be used as an alternative feed material for livestock. The nutritional content of cassava tubers is more complete than grass and legumes which are often used as animal feed for rural communities. The integration pattern between ruminants (cattle, buffalo, goat and sheep) with cassava plants can benefit cassava farmers. In addition, there can be reciprocal relationships with breeders so as to minimize vacant land. Empty land can be planted with cassava plants which can also serve as land for grazing for livestock. However, it should be understood that the cassava plant also has several drawbacks. One of the disadvantages is the presence of toxic compounds in cassava plants in the form of cyanide acid (HCN). The presence of this compound causes its use in poultry to be limited. Cassava plants contain two types of cyanogenic glycosides, namely a type of secondary metabolite in plants in the form of derivatives amino acids which are toxic. The types of cyanogenic glycosides in question are linamarin and small amounts of lotaustralin (methyl linamarin). Linamarin is rapidly hydrolyzed to glucose and acetone cyanohydrin while lotaustralin is hydrolyzed to cyanohydrin and glucose. Under neutral conditions, acetone cyanohydrin is decomposed to acetone and hydrogen cyanide. This cyanide acid causes the availability of the amino acid methionine in cassava to be low [60]. The content of cyanide acid compounds can be removed through a heating process. In addition, cassava plants, especially their leaves, are voluminous so they are not efficient in terms of transportation. Cassava is a wasteful plant in taking nutrients, so that many farmers are not interested in planting it.

The optimal use of cassava plants as ruminant feed can be done. One of them is by implementing an integration model of cassava plants with ruminant livestock cultivation (cattle, buffalo, goats and sheep). This model uses the principle that the livestock will come to the feed source. Some steps that can be taken include: (1) Livestock are raised around the cassava plantation. Efforts to use cassava plants in developing ruminant farms around or near the cassava planting area are carried out in the form of providing feed. This step is one way to take advantage of cassava plants, especially the leaves and tubers that have been rejected. The application of this system is carried out by raising livestock near cassava gardens which are then given by products from cassava plants such as leaves, rejected tubers and skin from the tubers. This system is quite beneficial because during the cultivation process of cassava plants, the manure produced by ruminants can then be used as organic fertilizer so that the fertility of the cassava plant land is maintained. For example, with the production of cassava leaves of 1.2 tons/ha, it is predicted to be able to meet the needs of 2 cows or buffaloes during the fattening period or the equivalent of 14 small ruminants (goats/sheep). The tuber part can be processed into tapioca products or chips. The waste products from cassava processing (skin and tuber) can be used as a mixture of animal feed ingredients. (2) Livestock are placed around the tapioca/chips factory area.

Every day, approximately tens of tons of cassava are produced into tapioca or chips. Furthermore, the waste generated from the production process in the form of skin and tuber can be used as a mixture of animal feed ingredients. Companies can use the land around the tapioca factory for the development of a livestock business. The company will have a very big opportunity to increase livestock production so that the meat needs of the Indonesian people can be met. The application of an integrated farm system through a combination of ruminant and non-ruminant livestock cultivation is thought to be more effective in optimizing the use of cassava plants. In order for this goal to be fulfilled, efforts that can be made are to raise livestock around the plantation. Through this integrated farm system, non-ruminant livestock (poultry, pig and horse) can be prioritized to utilize dried cassava and its leaves as a source of energy and protein. Meanwhile, the bark, leaves and cassava that have been rejected are used as a source of feed for ruminants (cattle, buffalo, goats and sheep). The system of integration of livestock in the main cassava plantation business has great opportunities and hopes because there are quite a lot of farmers and breeders engaged in this field.

Based on the results of economic analysis, the profit of a cassava farmer who owns land with an area of 1 ha is IDR 2,800,000 (assuming costs for planting a monoculture pattern for one growing season of 8 months). This means, the farmer only gets a profit of IDR 350,000 per month. This profit can be said to be inappropriate if used as the main business of a farmer. The value of profits obtained by a farmer will certainly increase if his business is integrated with livestock business (ruminants and/or non-ruminants). Competition in land use in the future as a consequence of efforts to maintain national food security and bioenergy development (bio-fuel) needs to be addressed immediately. Increasing productivity (intensification), especially on existing land, expanding the land area, and developing superior technological innovations are things that need to be done. Therefore, the use of vacant land for the development of an integrated system of plantations, agriculture and livestock is an appropriate and efficient innovation to do. This is an effort to find solutions to food security problems in Indonesia. This integration can increase food production in order to achieve the target food needs of the Indonesian people and increase the welfare of the community.

There is no doubt regarding the benefits of integrating cassava plants with livestock, for both farmers and breeders. The integration of plantation, agriculture and livestock crops is the best strategy in overcoming scarcity of resources in food production so that food can be fulfilled. With this integration, all agricultural activities can be economically profitable and ecologically sustainable. With the integrated pattern of plantations, agriculture and livestock, consumer demands for environmental sustainability, health and food safety, and the welfare of workers can be answered.

6. Integration of cattle with horticultural plants

Horticultural plants are one type of plant that is needed by humans to meet the needs of vitamins and minerals. Horticultural plants in the form of green vegetables have benefits as a source of vitamins and minerals that are important for fulfilling community nutrition. Increasing population, income and education will certainly affect public awareness of the importance of nutritional and health values.

Various efforts have been made by farmers so that horticultural crops can provide maximum results. Efforts that have been implemented include: (1) the use of superior variety seeds; (2) regular pest and disease control; (3) proper arrangement of spacing patterns; (4) timeliness and harvest time and (5) proper use of fertilizers. The application of fertilizers to horticultural crops needs attention, especially related to the type, dose, method and raw material (organic or inorganic). Continuous application of inorganic fertilizers can cause negative impacts on the environment. The impact that can arise is that the soil becomes solid due to the gluing effect, especially the type of ammonium fertilizer. The process of washing inorganic fertilizers in the soil can also have an effect on health when used as a source of drinking water. As an alternative, the use of organic fertilizers on land needs to be increased in availability for the balance of soil nutrients. Both types of fertilizers still need the right combination, because the availability of nutrients in organic fertilizers is not as complete as the nutrients in inorganic fertilizers [61]. Cow manure can reduce the cost of procuring fertilizers which at the same time can reduce production costs in addition to preserving soil organic matter, especially in sloped plantation areas [62].

The cultivation and development of cattle raised together with horticultural crops does not require new land resources. Horticultural crop waste can be used as animal feed that has been harvested so that the need for animal feed is always available every day. Farmers can make optimal use of vacant land to increase economic benefits. The input resources for livestock that are quite abundant, such as forage between plants (grass and legumes), can be directly used as a source of animal feed without disturbing the productivity of horticultural crops. Meanwhile, the potential for horticultural crop waste with simple technology can be used as a mixture of forage for cattle. The embankment area on horticultural plantations is still empty so it is still possible to use it as a superior grass cultivation area. One of the models applied in the implementation of the integration system between cattle and horticultural crops is the process of making fermented feed. The complete application of the model was presented in **Figure 4**.

With an integrated agricultural system, farmers will be more prosperous because there has been an increase in income. If the price of vegetables falls, the farmers still have other income, namely livestock. Cows can produce calves and fertilizer every year. Farmers who own cows use cow dung as raw material for biogas

Figure 4.
Utilization of vegetable waste as animal feed as a model for integrating cattle with horticultural crops. Source:: https://www.google.com/search?q=pemanfaatan+limbah+pasar+sebagai+pakan+sapi&tbm=isch&ved=2ahUKEwjpooj_rv3vAhWMUHoKHQoIAqQQ2-cCegQIABAA&oq=pemanfaatan+limbah+pasar+sebagai+pakan+sapi&gs_lcp=CgNpbWcQA1DI6AdY4P8HYKmJCGgAcAB4AIABkQGIAawKkgEDNC44mAEAoAEBqgELZ3dzLXdpei1pbWfAAQE&sclient=img&ei=e652YKmpF4yh9QOKkIigCg&bih=634&biw=1087&safe=strict&hl=id#imgrc=7qM2SDvdCt306M.

Description	Treatment	
	T_1	T_2
Initial Weight (IW) (kg)	269	270.42^{ns}
Final Weight (FW) (kg)	318.33	333.50^{*}
Average Daily Gain (ADG) (kg/day)	0.43	0.55^{*}
Consumption of Dry Matter (DM) (kg/day)	4.81	4.18^{ns}
Feed Cost Ratio (FCR)	11.19	7.69^{*}

Note: T_1 = 100% of grass + rice bran (1 kg/head/day); T_2 = 50% of grass + 50% of vegetable waste + rice bran (1 kg/head/day).
**Different superscripts on the same line indicate a marked difference.*
ns=no significant

Table 2.
Comparison of cattle productivity without the application of an integrated system (T_1) with the application of an integrated system with horticultural crops (vegetable waste) (T_2) at Antapan village, Baturiti district, Tabanan regency, Bali Province, Indonesia.

so they don't need to buy gas for cooking. An overview of the comparison of the productivity of cattle raised without the application of an integrated system with cattle raised using an integrated system with horticultural crops (vegetable waste) was presented in **Table 2.**

Based on the data in **Table 2**, it can be seen that the comparison of cattle groups that do not use an integrated system (T_1) has a lower average daily body weight (0.43 kg /day) than cattle raised in an integrated manner with horticultural crops (T_2) (0.55 kg/day). In addition, T2 treatment also has a lower FCR value (7.69) than T_1 (11.19). This shows that the T_2 treatment group is more efficient in utilizing feed to produce meat than T_1.

7. Conclusions

The integration model between plants and livestock or better known as integrated agriculture. This model combines the activities of the livestock industry with the agricultural industry. This model is often called a waste-free farming pattern because livestock waste is used as fertilizer for crops and agricultural waste is used as animal feed. The interaction between livestock and plants must be complementary, supportive and mutually beneficial so as to encourage an increase in the efficiency of profits from their farming.

Acknowledgements

The author would like to thank the government of the republic of Indonesia, in particular to the Ministry of Agriculture of the Republic of Indonesia, the Central Bureau of Statistics, the Ministry of Education and Culture for providing data as well as to all researchers and scientists who have contributed libraries and given permission to cite the results of their research.

Conflict of interest

The author hereby states that there is no conflict of interest between the author and other parties in the preparation of this article.

Author details

Muhammad Irfan Said
Faculty of Animal Science, Hasanuddin University, Makassar, Indonesia

*Address all correspondence to: irfanunhas@gmail.com; irfan.said@unhas.ac.id

References

[1] BPS. Statistical Yearbook of Indonesia 2020. BPS-Statistics Indonesia: Jakarta; 2020. ISSN 0126-2912

[2] Ministry of Agriculture. Agricultural Statistics. Ministry of Agriculture, Jakarta. [Internet]. 2020. Avalaible from: https://www.pertanian.go.id/ [Accessed: 2020-11-22]

[3] Anonymous. The need for beef is increasing. The Ministry of Agriculture will import. [Internet]. 2021. Avalaible from: https://www.republika.co.id/berita/qn9s7s423/kebutuhan-daging-sapi-meningkat-kementan-akan-impor [Accessed: 2021-04-15]

[4] Shadily, H. Indonesian Encyclopedia. Ichtiar Baru-Van Hoeve and Elsevier Publishing Projects: Jakarta; 1984. 2503 p

[5] FAO. Largest Rice Producer. [Internet]. 2011. Avalaible from: http://www.fao.org/faostat/en/#data; [Accessed: 2020-11-24]

[6] Balitbangtan. Superior Rice Varieties for the People Support Sustainable Rice Self-Sufficiency. Agricultural Research and Development Agency. Ministry of Agriculture: Jakarta; 68 p.

[7] Ministry of Agriculture. Roadmap to Increase National Rice Production (P2BN) towards a Rice Surplus of 10 Million Tons. Ministry of Agriculture: Jakarta; 2011. 40 p.

[8] Kusnadi U. Livestock technology innovation in the rice-livestock integration system to support beef self-sufficiency. Agricultural Innovation Development. 2008; 1(3): 189-205.

[9] Anonymous. The Predicted Demand for Beef in 2014 is 593,040 tons. [Internet]. 2013. Avalaible from: https://investor.id/agribusiness/ [Accessed 2020-11-24].

[10] Livestock Review. Meat Self-Sufficiency 2014: Hope or Dream. [Internet]. 2013. Avalaible from: http://livestockreview.com/swasembada-daging-2014-harapan-atau-mimpi-bag-i/ [Accessed: 2020-11-24]

[11] Livestock Review. If the Import Quota for Feeder Cattle is Reduced, The Breeders Will Benefit. [Internet]. 2013. Avalaible from: http://livestockreview.com/page/29/ [Accessed: 2020-11-24]

[12] Fagi AM, Ismail IG and Kartaatmadja S. Preliminary Evaluation of Crop-Livestock Farming Institutions In Several Districts in Central and East Java. In Fagi, A.M., and Hermanto (eds.). Plant-Animal Farming System. Agricultural Research and Development Agency. Ministry of Agriculture: Jakarta; 2004.

[13] Syam A, Sariubang M. Effect of organic fertilizers (cow dung compost) on rice productivity in irrigated land. In: Proceedings of the National Seminar on Plant-Animal Integration System. Livestock Research and Development Center and Casren; 20-24 July 2004: Denpasar, Bali; 93-103p.

[14] Sutardi, Musofie A, Soeharsono. Optimization of paddy production by utilizing organic fertilizers and integrated rice-livestock farming systems in lowland agro-ecosystems. In: Proceedings of the National Seminar on Plant-Animal Integration System. Livestock Research and Development Center and Casren; 20-24 July 2004: Denpasar, Bali; 1224-233 p.

[15] Haryanto B. The integrated system for livestock rice and cattle (SIPT) in the P3T program. In: Proceedings of the National Rice Week Seminar. Indonesian Rice Research Institute. 15-19 July 2004: Sukamandi.

[16] Nurawan A, Hadiana H, Sugandi D, Bachren S. The crop-livestock integration

farming system in Ciamis Regency, West Java. In: Proceedings of the National Seminar on Plant-Animal Integration System. Livestock Research and Development Center and Casren; 20-24 July 2004: Denpasar, Bali; 2004; p.133-141.

[17] Basuni R, Muladno, Kusmana C, Suryahadi. Beef cattle rice integration system in paddy fields. Food Crops Science and Technology Bulletin. 2010; 5(1): 31-48.

[18] Kariyasa, K. The crop-livestock integration system in the perspective of reorienting the fertilizer subsidy policy and increasing farmers' income. Analysis of Agricultural Policy. 2005; 3(1): 68-80.

[19] Swastika DKS. Building food independence and sovereignty to alleviate farmers from poverty. Agricultural Innovation Development. 2011; 4(2): 103-117.

[20] Sudirja R. Realizing food sovereignty through the policy of agri-food land management. In: Proceedings of Regional Seminar on Work Deliberation of the Executive Board of the Region II Indonesian Soil Science Student Association. Faculty of Agriculture, University of Padjajaran; Bandung; 2008. 8 p.

[21] Jenifer del RM. Module on Food Sovereignty. Malaysia: Pesticide Action Network Asia and the Pacific Press; 2007. 30 p.

[22] Arifin B. Building Food Independence and Sovereignty. Indonesian Science Congress: Jakarta; 2011. 24 p.

[23] Ika S. Food Sovereignty and Food Security: The state is obliged to make it happen: Fiscal Education Rubric; 2014. 12 p.

[24] Haryanto B, B. Harsana B, Ismet I. Integrated livestock rice system (SIPT). Technical Guidance. Balitnak: Bogor; 2003.

[25] Khairiah, Handoko DD. Fertilizer problems and the role of organic matter in the rice-cattle integration system in Perbaungan Sub-District, Serdang Bedagai Regency, North Sumatra. In : Proceedings of National Rice 2008. Center for Rice Research; Jakarta: 2009; p.1069-1076.

[26] Siswanto M, Patmawati NW, Trinayani NN, Wandia IN, Puja IK. Journal of Animal Science and Health. 2013; 1(1): 11-15.

[27] Basuni R, Muladno, Kusmana C, Suryahadi. Rice-beef cattle integration system model in paddy fields. Postgraduate Forum. 2010; 33(3): 177-190.

[28] Mariyono, Romjali E. Cheap Feed Innovation Technology For Beef Cattle Breeding Business. Center for Livestock Research and Development: Bogor; 2007.

[29] Prihartini I, Soebarinoto S, Chuzaemi, Winugroho M. Characteristics and degradation of fermented rice straw by lignolytic inoculum TLiD and BOpR. Journal of Animal Production. 2009; 11(1): 1-7.

[30] Sariubang M. The integrated farming system of Balinese cattle breeding with rice plants in paddy fields. Journal of Agrisistem. 2010; 6(1): 36-41.

[31] Abdulrachman S, Mejaya MJ, Sasmita P, Guswara A. Composting of Straw. Agricultural Research and Development Agency. Ministry of Agriculture: Jakarta; 2013.

[32] Haryanto B. Development of ruminant nutrition research. Wartazoa. 2012; 22(4):169-177.

[33] Yunilas. Rice Straw Biotechnology through Fermentation as a Feed

Ingredient for Ruminants. Scientific Work. Department of Animal Husbandry. Faculty of Agriculture. University of Northern Sumatra: Medan; 2009. 13 p.

[34] Ibrahim TM, Batubara LP, Simon P, Ginting, Doloksaribu M. Techniques for preserving and improving the quality of rice and corn straw using Starbio, Silage and Ammonia. Eds. Rusastra I.W. In: Proceedings of Site-Specific Technologies Towards Decentralization of Agricultural Development. Center for Agricultural Socio-Economic Research. Agricultural Research and Development Agency: Jakarta; 2000; p.397-401.

[35] Syamsu JA. Study on the Use of Microbial Starter in Rice Straw Fermentation as a Source of Feed on Smallholder Farms in Southeast Sulawesi. National Seminar on Biotechnology. Center for Biotechnology LIPI: Bogor; 2006.

[36] Suyasa IN, Guntoro S, Parwati, Suprapto, and Soethama IKW. The use of probiotics in the development of agribusiness-oriented beef cattle in Bali. Journal of the Assessment and Development of Agricultural Technology. Bogor Agricultural Socio-Economic Research Center: Bogor; 2004. 2(1).

[37] Tangendjaja B. The use of bran to make simple rations for laying ducks. In: Proceedings of the National Seminar on Livestock and the Second Poultry and Various Livestock Breeder Forum in Bogor. Ciawi Animal Research Institute; Bogor; 1988. 317-322 p.

[38] Udiyono. Possible use of rice bran as an ingredient for enzymes. In: Proceedings of Symposium on Bioprocess in Food Industry. 12-14 January 1987; Yogyakarta: 326-334 p.

[39] Bolsen KK, Ashbell G, Weinberg ZG. Silage fermentation and silage additives (Rev.). Asian-Australian J. of Anim. Sci. 1996; 9(5): 483-493.

[40] Salfina. Assessment of the provision of bran and probiotic bioplas to pregnant female DAS Katingan watershed local cows in Tumbang Liting Village, Katingan Hilir District, Central Kalimantan. Ziraa'ah. 2012; 33(1): 52-60.

[41] Muslim C, Nurasa T. Policy on beef cattle development in food crop-based production centers in Indonesia. Center for Socio-Economic Analysis and Agricultural Policy. Agricultural Research and Development Agency. Ministry of Agriculture: Jakarta; 2006. 13p.

[42] Muslim C. The Role of Beef Cattle Farmer Groups using the Rice-Livestock Integration System (RLIS) Approach in West Nusa Tenggara, East Java and West Java. Center for Socio-Economic Analysis and Agricultural Policy. Agricultural Research and Development Agency. Ministry of Agriculture: Jakarta; 2006. 17 p.

[43] Yuliani D. Animal rice integration system to achieve food sovereignty. J. Agrotechnology. 2014; 4(2): 15-26.

[44] Harli Integration system of goat-crops for resilient cocoa production. Agrovital. J. Agricultural Science. 2017; 2(1): 1-7.

[45] Wisri P, Susana IWR. The Potential and Utilization of Cocoa Fruit Skins as Alternative Feed for Ruminants. Wartazoa. 2014; 24(3): 151-159.

[46] Makka D. Prospects for the Development of a Competitive Livestock Integration System. Proceedings. National Seminar on Plant-Livestock Integration System. Center for Livestock Research and Development Collaborates with BPTP Bali and Casren: Denpasar; 2004. 18-31 p.

[47] Ickowicz A. Crop–Livestock Production Systems in the Sahel–Increasing Resilience for Adaptation to Climate Change and Preserving Food Security. Proceedings of a Joint FAO / OECD Workshop 23-24 April 2012.

[48] Director General of Livestock and Animal Health. Beef Cattle Population and National Meat Consumption in Indonesia. Director General of Livestock and Animal Health. Ministry of Agriculture: Jakarta; 2017.

[49] Umar S. Potential of Oil Palm Plantation as Center for Beef Cattle Development in Revitalizing and Accelerating Sustainable Livestock Development. Inauguration Speech for the Permanent Professor of Animal Reproduction at the Faculty of Agriculture. University of North Sumatra : Medan; 2009.

[50] Mathius IW, Sinurat AP, Tresnawati DP, Manurung BP. A study of Ready-to-Eat Feed Based on Oil Palm Industry By-Products for Pregnant Cows. [Internet]. 2007. Avalaible from: https://docplayer.info/33652661-Suatu-kajian-pakan-siap-saji-berbasis-produk-samping-industri-kelapa-sawit-untuk-sapi-bunting.html. [Accessed: 2020-11-24]

[51] Directorate of Animal Feed. General Guidelines for Animal Integration Development. Director General of Livestock and Animal Health. Ministry of Agriculture: Jakarta; 2011.

[52] Aritonang D. Oil palm plantations. Source of animal feed in Indonesia. J. Agricultural Research and Development Agency. 1986; 4: 93.

[53] Diwyanto K, Sitompul D, Manti I, Mathius I-W, Soentoro. Assessment of Business Development System Integration of Oil Palm-Cattle. In: Proceedings of the National Workshop on Oil Palm-Cattle Integration System. Ministry of Agriculture In cooperation with the Provincial Government of Bengkulu and PT. Agricinal: Jakarta; 2004.

[54] Setiawan BD, Arfa'i, Nur YS. Evaluation of the integrated Bali cattle breeding business management system with oil palm plantations in West Pasaman Regency, West Sumatra Province. Integrated Livestock Scientific Journal. 2019; 7(3): 276-286. DOI: http://dx.doi.org/10.23960/jipt.v7i3.p276-286

[55] Saragih B. Livestock Based Agribusiness. 2nd .Ed. USESE Foundation and Center for Development. Bogor Agricultural Institute: Bogor; 2000.

[56] Arfa'i. Potentials and Strategies for Developing Beef Cattle Business In Lima Puluh Kota Regency, West Sumatra. Dissertation. Postgraduate Program. Bogor Agricultural University: Bogor; 2009.

[57] FAO. World Agriculture. [Internet]. 2011. Avalaible from: http://www.world-agriculture.net/ [Accessed: 2020-11-24].

[58] BPS. Statistical Yearbook of Indonesia 2012. BPS-Statistics Indonesia: Jakarta; 2012.

[59] Radjit BS, Widodo Y, Saleh N, Prasetiaswati N. Technology to increase productivity and profits of cassava farming in ultisol dry land. Food Crop Science and Technology. 2014; 9(1): 51-62.

[60] Ramli N, Rismawati. Integration of cassava plants and poultry. Feed Commission Meeting Paper. 13-15 June 2007; Surabaya; 2007.

[61] Siswati L, Nizar R. Integrated agricultural model of horticulture and cows to increase farmers' income.

Indonesian Livestock Journal. 2012; 14(2): 379-384.

[62] Hutabarat TSPN. Regional Approach in Livestock Development. Directorate General of Livestock Production, Ministry of Agriculture: Jakarta; 2002. p.1-13.